Mathitude Quest
Middle School Math
Paula Hamilton

Mathitude, LLC
1031 Lucky Avenue
Menlo Park, CA 94025

info@mathitude.com

Cover design by Danielle Chritchley

Ordering Information:
Special discounts are available on quantity purchases by educators and others. For details, contact the publisher at the address above.

First Edition Mathitude Quest Middle School Math
First Printing, 2022
ISBN 978-1-7334698-7-6
www.mathitude.com

Mathitude Mahalos

I'll be forever indebted to our Mathitude 'auntie', Danielle Chritchley, who has been with me since the beginning of this journey and is integral to the heart, soul and vision that keeps us moving forward! Special thanks go out to Deanna S on the West Coast and Steve H on the East Coast. You've both been there every week for me and I've enjoyed every minute of our time together.

Many thanks, too, to Benji C and Aarthi P for your talent, creativity and hard work helping me to create this book. It's the first one I started writing and it's finally getting published...a long time coming.

And, with warmest alohas and mahalos, thank you to all the children and parents who have allowed me to be part of your mathematical journeys.

Let's keep striving for lifetime math engagement!

Paula Hamilton
May 2022

Table of Contents

STUDY SUGGESTIONS....7
WHOLE NUMBER OPERATIONS, CALCULATING AND ESTIMATING....9
WHOLE NUMBER OPERATIONS, BAR MODELING....10
FRACTION FRENZY WARM-UP 1....12
FRACTION FRENZY WARM-UP 2....13
FRACTION FRENZY, SUPER STRATEGIES....14
FRACTION FRENZY, ADVANCED PRACTICE....19
FRACTION FRENZY, MODELING AND VISUALIZATION....20
DECIMALS, MONEY MANIA....22
DECIMALS, POWERFUL PATTERNS....23
DECIMALS, DECIMAL DISCOVERY....25
ORDER, ORDER....26
AMAZING ALGEBRA....27
ALGEBRAIC TUG OF WAR....32
PROBABILITY PALOOZA....37
GEOMETRY CONCEPTS....40
RADICAL RATIOS....44
PERCENTS PLUS....47
DOMINO RATIOS....52
INTRO TO EXPONENTS....55
PRIME FACTORIZATION....57
PRIME FACTORIZATION, GCF AND LCM....59
DIVISIBILITY RULES....61
INTERESTING FRACTIONS....63
INTEGER INTRIGUE....65
GREENS VERSUS ALIENS....72
POSITIVELY NEGATIVITY....77
SUPER STATS....78
CROSS NUMBER....84
SECRET CODES....85
TOWER MATH....86
TRIANGLE CHALLENGES....87
MULTIPLICATION AND DIVISION QUEST....88

POWER PANEL: SOLVING QUIZZLES 89
DIVISIBILITY QUIZZLE 90
ANSWER KEY 91

Study Suggestions

Congratulations! You chose to engage with math and learn middle school math concepts early. Any time you choose to persevere and do quality work, you increase brain connections and make yourself a stronger mathematician. In this book, we'll give you study suggestions, review material you've already learned, introduce you to new material and provide lots of practice problems. Here's what you should keep in mind as you work through these exciting pages.

Suggestion 1 – Friendly Paper and Pencil

Paper and pencil are your friends. Many schools focus on mental math when you are younger. This helps you develop strong number sense and will be a good tool for you throughout your life. **For middle school math, the more you write down your work, the less silly errors you'll make.**

Suggestion 2 – Reading is Important

We'll give you many powerful processing tips in this book. In addition, we also introduce you to material you haven't seen yet. Be sure to read everything carefully – It may take you a couple times reading through to understand new material. This is also excellent practice for your middle school classes.

Suggestion 3 – Study Your Errors

Study your errors. It is more important to study your errors than it is to do large amounts of problem sets. When you discover an error as you work, do the following:

- Circle any mistakes you make in a nice bright color.
- Write down why you made the mistake (in the same nice bright color).
- Practice similar problems so you can figure out how to avoid making that mistake in the future.

Suggestion 4 – Take What You Need

This book provides more instruction and exercises than anyone could do in three months. Try a few exercises from each page/topic and then move onto the next topic. Once you've seen all the topics, do more work in the areas where you feel you need more practice.

Suggestion 5 – Develop a Study Schedule

Develop a study schedule. Decide (with your parents) how much time you will spend on these pages each day. If you're like us, you may find it hard to stop!

Suggestion 6 – Store Work in a Notebook or Binder

Set up a notebook or binder where you store all your work for this quest. We didn't provide working space on our pages and you need to keep your work to study your errors. Be sure to label the work you are doing…What page are you working on? What question is it? Have you boxed or circled your answer so it is easy to check your work?

SUGGESTION 7 – ASK FOR HELP
As you work through this material, there will be times when you don't understand a concept or need to figure out how to stop making the same error. Be sure to ask your parents, or friends, for help if you get stuck. You can also email us at info@mathitude.com.

SUGGESTION 8 – EXPERIMENT, FAIL, LEARN, REPEAT
Don't be discouraged if you make errors when you work through this math quest. When you first learned to walk you fell down a lot, but we bet you are an excellent walker now. It's the same thing with any skill you are working on mastering. You try it, you don't get it quite right, you think about what you'll do differently the next time, and then you try again.

SUGGESTION 9 – HEAVY LIFTING GETS YOU READY FOR EASIER WORK
Some of the materials in this guide are designed to be very challenging. We expect you to struggle, not get the right answer the first time and to learn some middle school math before you've seen it at school. You can assume that most of the material will be challenging.

SUGGESTION 10 – BRING ON THE FUN
We've provided bonus materials at the end that are quite fun. You could probably do some of the work on those pages now but you'll need to work through the entire quest book in order to solve all of these puzzles. In addition, as you improve your math skills and knowledge, you will discover a vastly entertaining and interesting world full of awe-inspiring mathematical concepts.

SUGGESTION 11 –BREATHE!
While we know that committing to a math goal can be scary, make sure you remind yourself, and your parents, that everything will be okay. You will do your best mastering this material, you will learn more math and you will learn more about how your brain works. Enjoy the process and you'll never need to worry about the results.

Let's Get Started!

Middle School Math Quest

Whole Number Operations, Calculating and Estimating

Let's warm up with calculation, rounding and modeling exercises.

Multiply.

1. 421×357

2. 689×807

3. 805×742

Divide. Answers should be left as whole numbers with a remainder (you don't need to practice decimal division for these problems).

4. $8{,}222 \div 96$

5. $6{,}759 \div 62$

6. $56{,}302 \div 74$

Round each number to the place values shown in the table. The first table has been completed for you.

EXAMPLE

ROUND TO THE NEAREST...	NUMBER 26,901
TENS	26,900
HUNDREDS	26,900
THOUSANDS	27,000
TEN THOUSANDS	30,000
HUNDRED THOUSANDS	0

Q7

ROUND TO THE NEAREST...	NUMBER 59,261
TENS	
HUNDREDS	
THOUSANDS	
TEN THOUSANDS	
HUNDRED THOUSANDS	

Q8

ROUND TO THE NEAREST...	NUMBER 167,089
TENS	
HUNDREDS	
THOUSANDS	
TEN THOUSANDS	
HUNDRED THOUSANDS	

Draw a line from each expression to its best estimate. Round your numbers BEFORE you do your operations.

EXPRESSION	**ESTIMATE**
9. $467 + 51$	
	500
10. $23{,}849 \div 442$	
11. 45×12	
	600
12. $647 - 157$	
13. $642{,}345 \div 982$	Other Estimate
14. $237 \times 29 \div 11$	

Middle School Math Quest

Whole Number Operations, Bar Modeling

Solve each of these challenges. Use bar modeling (or other modeling techniques) to help you get the right answer. Show all your work and don't forget to include labels in your answer (books, pencils, etc.)

Challenge 1: Logan has three times as many books as her brother. If they have 60 books in all, how many books does Logan have?

Challenge 2: If Willem has 20 more pencils than Hannah and Hannah has 55 less pencils than Jackson, how many pencils does Jackson have if they have 150 pencils altogether?

Challenge 3: Grace has twice as many apples as Benji. If there are 72 apples in total, how many apples does Grace have?

Challenge 4: Alice has twice as many fidget spinners as Mark. Mark has 3 less fidget spinners than Brandon. If they have a total of 307 fidget spinners, how many does each child have?

Challenge 5: There are 50 students on the bus for the field trip to Raging Waters. If there are 10 more boys than girls, how many boys and girls are on the bus?

Challenge 6: There are 380 passengers on the plane. There are 70 more adults than there are children. How many adults are on the flight?

Challenge 7: Three consecutive numbers sum to 354. What are the numbers?

Challenge 8: Four consecutive odd numbers sum to 256. What are the numbers?

Challenge 9: Suzanne has $1,200 remaining from her paycheck. One-fourth of that amount will be put away for savings. The remaining amount will be divided among her four children's allowances. The oldest child gets four times as much money as the youngest and the two middle children each receive half as much as the oldest. How much does each child get for their monthly allowance?

Challenge 10: On Monday, Ben has twice as much money as Jerry. During the week, they both earn $50. At that point, Ben has $75 more than Jerry. How much money did they both start with on Monday?

Challenge 11: Jack has five times as many candies as his brother. If they have 66 candies in all, how many candies does Jack have?

Challenge 12: Sam has three times as many apples as Finn. If there are 84 apples in total, how many apples does Finn have?

Challenge 13: There are 98 students on the bus to Outdoor Ed. If there are 6 more boys than girls, how many boys and how many girls are on the bus?

Challenge 14: Three consecutive numbers sum to 468. What are the numbers?

Challenge 15: On Monday, Kat has twice as much money as Bobby. During the week, they both earn $65. At that point, Kat has $20 more than Bobby. How much money did they both start with on Monday?

Challenge 16: Angelina scored 2 more points than John and John scored 1 more point than Sarah. How many points did Sarah score if they scored 28 points in total?

Challenge 17: Bella has twice as many marbles as Milo. Milo has 6 less marbles than Ron. If they have a total of 82 marbles, how many marbles does each child have?

Challenge 18: There are 360 people on the school trip. There are two more fifth graders than fourth graders. How many fifth graders are there?

Middle School Math Quest

Fraction Frenzy Warm-Up 1

Let's master basic fraction operations. All answers must be in simplest form.

1. $\frac{2}{6}+\frac{3}{6}=$

2. $\frac{1}{3}+\frac{4}{5}=$

3. $\frac{8}{9}-\frac{1}{2}=$

4. $2\frac{1}{2}+1\frac{3}{4}=$

5. $1\frac{2}{3}+4\frac{4}{9}=$

6. $6+\frac{1}{3}=$

7. $4\frac{1}{8}-2\frac{5}{6}=$

8. $6\frac{3}{4}-2\frac{1}{3}=$

9. $6-\frac{5}{3}=$

10. $\frac{3}{24}\times\frac{4}{16}=$

11. $4\times\frac{3}{5}=$

12. $2\frac{3}{4}\times 4=$

13. $\frac{3}{8}\times\frac{8}{27}=$

14. $10\div\frac{1}{4}=$

15. $\frac{5}{8}\div 10=$

16. $\frac{15}{9}+\frac{5}{4}=$

17. $\frac{16}{9}\times\frac{12}{8}=$

18. $\frac{20}{8}\div 4=$

Middle School Math Quest

Fraction Frenzy Warm-Up 2

If you were a little rusty with your first page of fraction practice, complete this practice, too. Don't forget – All answers must be written in simplest form.

1. $\frac{2}{7}+\frac{3}{7}=$
2. $\frac{3}{5}+\frac{2}{3}=$
3. $\frac{2}{3}-\frac{3}{6}=$
4. $2\frac{1}{2}+3\frac{2}{4}=$
5. $3\frac{1}{3}+1\frac{2}{6}=$
6. $3+\frac{3}{4}=$
7. $4\frac{1}{2}-3\frac{2}{3}=$
8. $6\frac{1}{2}-2\frac{3}{4}=$
9. $4-\frac{3}{2}=$
10. $\frac{4}{8}\times\frac{6}{15}=$
11. $3\times\frac{2}{3}=$
12. $4\frac{1}{3}\times 3=$
13. $\frac{2}{6}\times\frac{4}{10}=$
14. $6\div\frac{1}{2}=$
15. $\frac{3}{8}\div 6=$
16. $\frac{12}{9}+\frac{10}{4}=$
17. $\frac{15}{9}\times\frac{6}{10}=$
18. $\frac{15}{12}\div 3=$

Middle School Math Quest

Fraction Frenzy, Super Strategies

In the next section, we'll give you middle school fraction practice. Before we do that, we want to show you strategies for working with fractions. Pay attention to these – They'll help you avoid many silly errors and will simplify your calculations.

Fraction Super Strategy 1 - Stack It

STRATEGY 1 – STACK IT: When you add or subtract mixed numbers, stack the numbers and use the space to the right to do additional calculations.

Example: $14\frac{1}{3} - 5\frac{7}{9}$

Stack It →	Common Denominators →	ReGroup →	Add/Subtract
$14\frac{1}{3}$	$14\frac{\not 1}{\not 3}\ \frac{4}{12}$	$\overset{13}{\not{14}}\frac{4}{12} + \frac{12}{12}$	$13\frac{16}{12}$
$-\ 5\frac{3}{4}$	$-\ 5\frac{\not 3}{\not 4}\ \frac{9}{12}$	$-\ 5\frac{9}{12}$	$-\ 5\frac{9}{12}$
			$8\frac{7}{12}$

Practice **Stack It** with these problems:

1. $27\frac{1}{5} - 13\frac{7}{15}$
2. $21\frac{2}{9} - 19\frac{5}{6}$
3. $16\frac{3}{4} + 6\frac{7}{8}$
4. $17\frac{1}{5} + 4\frac{1}{2}$
5. $16\frac{2}{3} - 7\frac{1}{2}$
6. $28\frac{3}{5} - 4\frac{7}{10}$

Keep on Stacking!

Fraction Super Strategy 2 – Improper is Proper

Strategy 2 - Improper is Proper: For multiplication and division ONLY, change mixed numbers to improper fractions before you multiply or divide.

Example: $1\frac{1}{5} \times 3\frac{2}{7}$

Mixed Number to Improper Fraction → **Multiply Numerators and Denominators** → **Improper Fraction or Mixed Number Answer**

$\frac{6}{5} \times \frac{23}{7}$ → $\frac{6 \times 23}{5 \times 7} = \frac{138}{35}$ → $\frac{138}{35}$ or $3\frac{33}{35}$

You can leave your answers as improper fractions if the numerator and denominator don't have common factors.

Practice **Improper is Proper** with these problems:

7. $2\frac{3}{8} \times 3\frac{4}{5}$

8. $1\frac{1}{11} \times 1\frac{1}{5}$

9. $2\frac{1}{3} \times 4\frac{3}{4}$

10. $1\frac{4}{5} \times 3\frac{1}{2}$

11. $1\frac{2}{3} \times 8\frac{7}{4}$

12. $12\frac{1}{4} \times 7\frac{2}{3}$

Middle School Math Quest

Fraction Super Strategy 3 - Simplify Before You Multiply

STRATEGY 3 - SIMPLIFY BEFORE YOU MULTIPLY: When multiplying two or more fractions together, you should simplify numerators and denominators before you multiply – It keeps your calculations simple.

Find common factors between numerators and denominators – You can simplify across fractions (Example 1) or simplify within a fraction (Example 2).

Example 1: $\frac{36}{39} \times \frac{13}{9}$

$36 \div 9 = 4$ $\quad 13 \div 13 = 1$

$$\frac{\cancel{36}}{\cancel{39}} \times \frac{\cancel{13}}{\cancel{9}} = \frac{4 \times 1}{3 \times 1} = \frac{4}{3}$$

$39 \div 13 = 3$ $\quad 9 \div 9 = 1$

Example 2: $\frac{4}{8} \times \frac{3}{15}$

$4 \div 4 = 1$ $\quad 3 \div 3 = 1$

$$\frac{\cancel{4}}{\cancel{8}} \times \frac{\cancel{3}}{\cancel{15}} = \frac{1 \times 1}{2 \times 5} = \frac{1}{10}$$

$8 \div 4 = 2$ $\quad 15 \div 3 = 5$

Practice **Simplify Before You Multiply** with these problems:

13. $\frac{16}{24} \times \frac{48}{64}$

14. $1\frac{2}{5} \times \frac{10}{14}$

15. $3\frac{3}{4} \times 4\frac{2}{3}$

16. $\frac{15}{20} \times \frac{72}{168}$

17. $1\frac{5}{7} \times \frac{27}{63}$

18. $2\frac{16}{36} \times 5\frac{4}{6}$

Fraction Super Strategy 4 – Keep, Change, Flip

STRATEGY 4 - KEEP, CHANGE, FLIP (KCF): Remember the algorithm for fraction division by using Keep, Change, Flip – Keep the first number as is, Change the division operation to multiplication, Flip the divisor (reciprocal of the number).

Example: $2\frac{3}{4} \div 1\frac{1}{6} = \frac{11}{4} \div \frac{7}{6} = \frac{11}{4} \times \frac{6}{7} = \frac{11}{2} \times \frac{3}{7} = \frac{33}{14}$

Convert mixed numbers to improper fractions

Keep Change Flip

Simplify Before You Multiply

Practice **Keep, Change, Flip** with these problems:

19. $3\frac{1}{3} \div \frac{1}{3}$

20. $\frac{4}{5} \div \frac{2}{3}$

21. $3\frac{12}{18} \div \frac{10}{12}$

22. $3\frac{2}{3} \div \frac{5}{7}$

23. $1\frac{5}{6} \div \frac{4}{3}$

24. $7\frac{1}{4} \div 6\frac{1}{5}$

Middle School Math Quest

Fraction Super Strategy 5 – Keep It Simple, Smartie!

STRATEGY 5 - KEEP IT SIMPLE, SMARTIE (KISS): Always remember to simplify fractions before submitting as final answers. You'll give away many points if you don't simplify your answer.

Practice **KISS** by changing these fractions to their simplest form representation:

25. $\frac{57}{81}$

26. $\frac{64}{72}$

27. $\frac{60}{96}$

28. $\frac{72}{96}$

29. $\frac{24}{30}$

30. $\frac{147}{763}$

31. $\frac{15}{165}$

32. $\frac{96}{60}$

33. $\frac{555}{110}$

34. $\frac{46}{68}$

35. $\frac{10{,}200}{510}$

36. $\frac{85}{204}$

Middle School Math Quest

Fraction Frenzy, Advanced Practice

Use your super strategies to solve these tougher fraction problems. Don't forget to simplify your answers. Improper fractions or mixed number answers are fine – Just make sure the fraction portion of your answer is in simplest form.

1. $\frac{12}{32} \times \frac{16}{18} =$

2. $\frac{12}{20} \times \frac{7}{16} =$

3. $\frac{7}{8} \div \frac{42}{64} =$

4. $\frac{11}{12} \div \frac{3}{11} =$

5. $32\frac{5}{6} + 29\frac{7}{8} =$

6. $19\frac{1}{9} - 12\frac{3}{4} =$

7. $18\frac{1}{3} - 12\frac{5}{6} =$

8. $47\frac{1}{8} - 29\frac{2}{3} =$

9. $25\frac{4}{5} + 18\frac{7}{10} =$

10. $20 \div \frac{32}{40} =$

11. $8\frac{3}{4} \div 4\frac{7}{14} =$

12. $11\frac{2}{5} \times 2\frac{3}{4} =$

13. $12\frac{1}{5} \times 3\frac{1}{4} =$

14. $11\frac{1}{6} - 8\frac{1}{4} =$

15. $59\frac{2}{5} + 18\frac{9}{10} =$

16. $37\frac{5}{8} - 19\frac{5}{6} =$

17. $50 \div \frac{25}{30} =$

18. $10\frac{2}{4} \div 5\frac{6}{8} =$

19. $\frac{25}{40} \times \frac{28}{35} =$

20. $6\frac{5}{12} - \frac{3}{4} =$

21. $\frac{11}{15} \div 3\frac{2}{3} =$

22. $\frac{90}{108} \div \frac{40}{48} =$

23. $1\frac{5}{9} + 3\frac{7}{6} =$

24. $\frac{24}{32} \times \frac{8}{56} =$

Middle School Math Quest

Fraction Frenzy, Modeling and Visualization

You've practiced many of the algorithms that will help you master middle school fractions. Now, let's practice modeling fraction calculations. Visualization will give you a much deeper understanding of fractions.

Use bar modeling to solve these problems:

1. Thomas made $3\frac{1}{3}$ quarts of lemonade and Ava prepared four times as much as Thomas. If they need to fill a 10-gallon container, how much more lemonade do they need to make?

2. Emily sells chocolates as part of a school event. She sold 3 the first hour, then $\frac{2}{3}$ of that amount the second hour. In her third hour of selling, she sold double the amount she sold the first hour. How many chocolates did she sell in total?

3. Ellis buys 4 sets of earrings. Her mom buys 3 more than $\frac{1}{2}$ that number, and her sister buys twice as many as her mother. How many pairs of earrings do they buy in total?

4. Steve buys 48 hats to sell. $\frac{1}{4}$ are red, $\frac{1}{3}$ are blue, and half of the remaining are purple. The rest of the hats are black. How many black hats does he have?

5. The local football stadium complex covers $11\frac{2}{3}$ acres of land. If $\frac{2}{7}$ of the land is reserved for the parking lot, how much of the land is remaining for the stadium facilities?

6. Peter sold $\frac{1}{2}$ of his school raffle tickets on Monday. Then, on Tuesday, he sold $\frac{1}{3}$ of his remaining tickets. Now, he has 16 tickets remaining. How many tickets did he start with on Monday?

Show models to represent these problems:

7. $3 \div \frac{1}{4}$

8. $\frac{5}{6} \times \frac{1}{2}$

9. $\frac{5}{7} \div 2$

10. Write a word problem that could match this expression: $2\frac{2}{3} \div \frac{1}{6}$. Then, solve it.

Decide whether these statements are *always, sometimes* or *never* true. If your answer is *sometimes*, provide examples where the statement is true and not true. If your answer is *always* or *never*, provide two examples.

11. A number n divided by an improper fraction is less than n.

12. A number x multiplied by another number is greater than x.

13. When you divide a number y by a fraction less than 1, your answer will be smaller than y.

14. When you divide the number 10 by a fraction, your answer will be greater than 10.

15. Multiplying a number x by $\frac{1}{2}$ gives an answer larger than x.

16. When you multiply two proper fractions, the answer will be less than one.

Provide true examples for each of these questions.

17. A fraction multiplied by a mixed number gives an answer greater than the mixed number.

18. A number divided by an improper fraction gives an answer less than 1.

19. A mixed number multiplied by a proper fraction is less than 1.

"You really understand fractions!"

Middle School Math Quest

Decimals, Money Mania

Fitz's Candy Store needs your help. His cash register and calculators are all broken, and he needs to know how much each customer owes him. Can you calculate the total each customer owes? Fitz calculated the first one for you.

CANDY STORE PRICES

Candy	Price per lb
Lemon Bees	\$2.25
Chocolate Bombs	50 ¢
Pizza Gummies	\$1.50
Sour Fizzes	\$1.00

Customer 1

$\frac{2}{5}$ lb Choc Bombs

$= \frac{2}{5} \times .50 = .4 \times .5 = \$.20$

$3\frac{1}{2}$ lbs Sour Fizzes

$= 3.5 \times 1 = \$3.50$

$\frac{1}{2}$ lb Pizza Gummies

$= .5 \times 1.5 = \$0.75$

TOTAL = \$4.45

Customer 2

4 lbs Choc Bombs

$\frac{7}{10}$ lb Pizza Gummies

2 lbs Lemon Bees

$3\frac{1}{2}$ lbs Sour Fizzes

TOTAL =

Customer 3

$\frac{1}{5}$ lb Choc Bombs

$\frac{1}{3}$ lb Pizza Gummies

3 lbs Lemon Bees

TOTAL =

Customer 4

$\frac{3}{4}$ lb Sour Fizzes

$4\frac{1}{2}$ lbs Pizza Gummies

2 lbs Lemon Bees

TOTAL =

Customer 5

$\frac{2}{3}$ lb Pizza Gummies

$1\frac{2}{3}$ lb Lemon Bees

$\frac{1}{4}$ lb Sour Fizzes

TOTAL =

Customer 6

$4\frac{3}{5}$ lbs Sour Fizzes

$2\frac{1}{6}$ lbs Pizza Gummies

TOTAL =

Customer 7

$3\frac{1}{2}$ lbs Choc Bombs

$2\frac{4}{9}$ lbs Lemon Bees

$1\frac{5}{6}$ lbs Pizza Gummies

$2\frac{3}{10}$ lbs Sour Fizzes

TOTAL =

Customer 8

10 lbs Lemon Bees

10 lbs Choc Bombs

10 lbs Pizza Gummies

TOTAL =

Customer 9

5 lbs Lemon Bees

3 lbs Pizza Gummies

4 lbs Sour Fizzes

$5\frac{7}{10}$ lbs Choc Bombs

TOTAL =

Middle School Math Quest

Decimals, Powerful Patterns

If you understand the number patterns related to decimal multiplication and division, you'll be able to handle many complicated questions. Learn and practice how to work with these decimal patterns.

Decimal Explorations, Recognizing Patterns

Example 1: 34 x 100

34 x 100 is 34 x 1 Hundred
34 x 1 Hundred is 34 Hundreds
34 Hundreds is 3,400

Conjecture: When we multiply a number by 100, we move the decimal point two places to the right

Example 2: 34 x 200

34 x 200 is 34 x 2 Hundreds
34 x 2 Hundreds is 68 Hundreds
68 Hundreds is 6,800

Conjecture: When you multiply a number by a multiple of 100, first multiply the number (34) by the multiple of 100 (2) and then move the decimal point two places to the right

Example 3: 34 x 0.1

$$34 \times 0.1 = 34 \times \frac{1}{10} = 34 \div 10 = 3.4$$

Conjecture: When you multiply a number by 0.1, you move the decimal point one place to the left.

You can use fraction operations to help you understand decimal operations.

Example 4: 0.34 x 1,000

34 x 1 = 34
34 x 10 = 340
34 x 100 = 3,400
34 x 1,000 = 34,000
3.4 x 1,000 = 3,400 (3.4 is 10 times smaller than 34)
0.34 x 1,000 = 340

Conjecture: Start with simpler numbers and use a **sequence approach** to solve harder decimal problems.

Make sure you understand each step in the examples above before you move onto the remaining decimal pages.

Fill in the blank spaces in the table below. This will help you with the exercises on the following page.

Decimal Patterns

When You...	A Number By...	Move the Decimal Point....	Place(s) to the...	Is Answer Smaller or Larger Than Original Number?
Multiply		One	Right	Larger
Multiply	100	Two		Larger
Multiply	0.1		Left	Smaller
Multiply	0.01	Two		Smaller
Divide	10	One	Left	
Divide		Two	Left	Smaller
Divide	0.1	One	Right	Larger
Divide	0.01	Two	Right	

Middle School Math Quest

Decimals, Decimal Discovery

Now that you've learned decimal patterns, find the answers to these questions. Use fraction operations, bar models or the sequence method to support your final solutions. You won't need to use standard multiplication and division algorithms if you think about the decimal patterns you've explored.

1.	6	×	4	=	0.6	×	0.4	=	0.6	×	0.04	=
2.	6	×	400	=	60	×	400	=	600	×	400	=
3.	6	×	0.04	=	60	×	0.004	=	600	×	0.04	=
4.	24	÷	6	=	2.4	÷	6	=	0.24	÷	6	=
5.	240	÷	6	=	2,400	÷	60	=	2,400	÷	600	=
6.	2.4	÷	60	=	2.4	÷	600	=	0.24	÷	60	=
7.	24	÷	0.4	=	2.4	÷	0.4	=	0.24	÷	0.4	=
8.	2,400	÷	0.4	=	240	÷	0.04	=	2,400	÷	0.004	=
9.	2	×	90	=	1.8	÷	0.09	=	200	×	0.9	=
10.	800	×	0.05	=	400	÷	50	=	0.4	÷	0.5	=
11.	630	÷	0.07	=	7	×	0.009	=	0.63	÷	9	=
12.	36	÷	0.9	=	90	×	0.4	=	900	×	0.4	=

Middle School Math Quest

Order, Order

Order has a couple different meanings in elementary and middle school math. We talk about ordering numbers from greatest to smallest or smallest to greatest. We also talk about order of operations. Order of operations helps mathematicians know that a person, calculator or computer will always calculate an expression in the same order and get the same answer as other mathematicians. We'll practice both types of ordering on this sheet.

Order these groups from greatest to least

Group 1: $0.62, 0.6095, 6.01, 6.003, 6.0035$

Group 2: $1\frac{4}{5}, 2\frac{1}{2}, 1\frac{2}{3}, \frac{3}{5}, 2\frac{1}{6}$

Group 3: $2\frac{1}{2}, 2.46, 1\frac{3}{5}, 2.12, 2\frac{7}{9}$

Group 4: $2.014, 2\frac{3}{4}, 2.14, 2\frac{1}{3}, 2.102$

Practice these order of operations exercises. Remember....

- When you only have addition and subtraction, work from left to right.
- When you only have multiplication and division, work from left to right.
- When you have mixed operations, follow PEMDAS (Parentheses then Exponents then Multiplication & Division then Addition & Subtraction).
- Only do one operation at a time so you don't make silly errors.

We've done the first one for you.

1. $16 + 3 \times 7$ $= 16 + 21$ (Multiplication before addition) $= 37$	6. $10\frac{3}{4} - 5\frac{3}{4} \times \frac{1}{2} + 3$
2. $11 - 1 + 1 \times 1 - 1 + 111 \times (1 + 1 + 1 - 1)$	7. $\frac{1}{2} + \frac{1}{2} \times \frac{1}{2} - \frac{1}{2} \times \left(\frac{1}{2} + \frac{1}{2} \times \frac{1}{2} - \frac{1}{2}\right)$
3. $72 \div 2 + 3 \times 5$	8. $4\frac{3}{4} - \frac{7}{8} \div \frac{1}{4}$
4. $20 - 8 \div 4 \times 2$	9. $3 \div \frac{2}{3} - \frac{1}{3} \div \frac{1}{6}$
5. $67 - (37 - 17) \times 2$	10. $\frac{1}{4} \div \frac{1}{4} \div \frac{1}{4} - \frac{1}{4} \times \frac{3}{4}$

Amazing Algebra

You may not know it but you've been doing algebra since 1st grade.

Algebra uses letters to represent what we don't know. Up until now, your textbooks have used blank spaces (like in Example 1), bar models or words to represent what we don't know. Now, we'll use letters (like x, y and z) to represent those question marks, boxes and underlines.

Algebraic Expressions

Without Algebra

Example 1

_______ + 12 = 20

Example 2

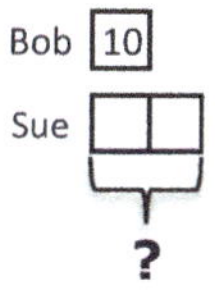

Example 3

Jimmy has 3 more cookies than Ellen. How many cookies does Jimmy have?

Convert these examples to algebra. When an expression or equation includes letters, it's called an **algebraic expression (or equation).**

With Algebra

Example 1

Our unknown is the number we're adding to 12.

$x + 12 = 20$

Example 2

Our unknown is how much Sue has.

$x \div 2 = 10$

Example 3

Our unknown is how many cookies Ellen has.

$x + 3$

Practice changing words into algebraic expressions with these exercises. The first one is done for you.

1. 3 less than a number n

   $n - 3$

2. Two times more than y

3. 6 more than x

4. Half of a number z

5. 5 less than half of x

6. 4 more than three times b

When we have an algebraic expression and are given a value for the variable, we can evaluate the expression.

Evaluating Algebraic Expressions

Example	**Example**
Algebraic Expression: $x \div 2 + 3$	Algebraic Expression: $x + 3y - 7$ $x = 6$ and $y = 4$
$x = 10$	Note: $3y$ means $y + y + y$ (The 3 means "add y together 3 times or multiply y by 3).
Evaluate the expression by replacing the x in the expression with 10.	
$10 \div 2 + 3 = 5 + 3 = 8$	Evaluate: $6 + 3 \times 4 - 7 = 6 + 12 - 7 = 11$
$\mathbf{x \div 2 + 3 = 8}$ **when** $\mathbf{x = 10}$	$\mathbf{x + 3y - 7 = 11}$ **when** $\mathbf{x = 6}$ **and** $\mathbf{y = 4}$.

Practice evaluating these algebraic expressions using the given values:

7. $3x - 20$ when $x = 16$

8. $5x + 2y$ when $x = 2$ and $y = 5$

9. $\frac{x}{7} - 4$ when $x = 63$

Keep on Going!

Algebraic expressions include **variables** (letters like x and y) and numbers. We call two terms **like terms** if they represent the same thing. **Like terms** can be combined to simplify your algebraic expressions.

Simplifying Algebraic Expressions

Example 1

Expression: $3x + 5x - 4$

Coefficients – They tell us how many x's to add together.

$3x$ and $5x$ are **like terms** because they represent the same unknown value.

The numbers in front of our variables (3 and 5) are called the **coefficients**. They tell us how many of each variable to add together.

Simplified Expression:

$3x + 5x - 4 = 8x - 4$

*Since the 3 coefficient tells us to add 3 x's together and the 5 coefficient tells us to add 5 x's together, we have 8 x's in all.

Example 2

Expression: $7x + 3y - 6x + 4 - 3$

Constants – Stand alone numbers

$7x$ and $6x$ are like terms.
4 and 3 are like terms.
$3y$ doesn't have like terms in this expression. y represents a different unknown value than x.

We can combine our like terms...

Simplified Expression:

$7x + 3y - 6x + 4 - 3 =$
$7x - 6x + 3y + 4 - 3 = x + 3y + 1$

Practice simplifying these algebraic expressions by combining like terms:

10. $x + 3x + 7 - 6$

11. $5y + 6 - 3y + 2y - 3 + 2$

12. $14 + 10x - 7 + 4x - 4$

13. $5x + 7y + 3 - 4x - 6y - 2$

14. $13 + 4x - 7 + 3y - 2x - 4 - 2x - y - 2 - 2y$

So far, we've mainly worked with algebraic expressions. Now, we'll show you strategies you can use to solve algebraic equations for unknown variables.

Algebraic Solving Strategies

You can use these powerful strategies to help you figure out when the equation will be true. It will only be true for one value of the variable.

Addition Property of Equality: When you add the same number to both sides of an equation, the equation is still true.

Subtraction Property of Equality: When you subtract the same number from both sides of an equation, the equation is still true.

ADDITION PROPERTY
Bar Modeling Example

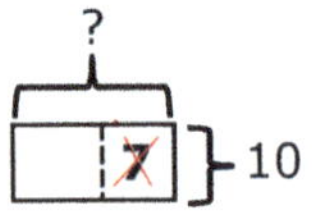

Bar model shows us an unknown number. When we subtract 7 from that number, we make 10.

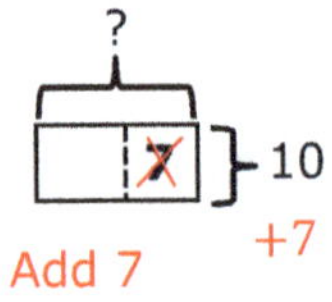

We can add 7 to the bar model and to the number 10 using the addition property of equality.

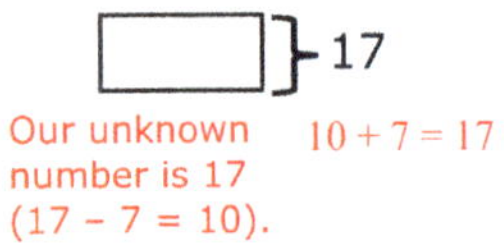

Simplify the model by combining the number terms.

Our unknown number is 17. When you start with 17 and take away 7, you're left with 10.

SUBTRACTION PROPERTY
Bar Modeling Example

Bar model shows us when you add 10 to an unknown number, you make 28.

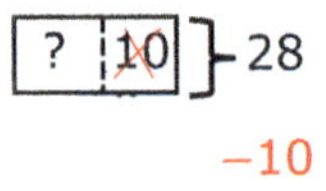

We can subtract 10 from the bar model and also from 28 using the subtraction property of equality.

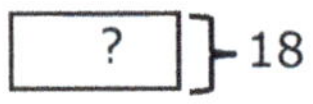

Simplify the model.

Our unknown number is 18. When you add 18 and 10, you make 28.

Algebra Example

$$x - 7 = 10$$
$$+7 \qquad +7$$

← Add 7 to both sides

$$x - 7 + 7 = 10 + 7$$
$$x = \mathbf{17}$$

← Combine like terms and simplify.

Algebra Example

$$10 + x = 28$$

$$10 - 10 + x = 28 - 10$$
$$x = \mathbf{18}$$

← Combine like terms and simplify.

The **Multiplication Property of Equality** and **Division Property of Equality** use the same solving methods.

Middle School Math Quest

Solve these algebraic equations for the unknown variable using addition/subtraction properties of equality.

15. $32 - x = 22$

16. $x + 30 = 30$

17. $x - 20 = 20$

18. $37 + x = 60$

19. $x + x + 10 = 40$

20. $2x - 3 = 25$

21. $x - 243 = 176$

22. $13 + x = 1{,}000$

23. $x + \frac{1}{4} = \frac{3}{8}$

24. $12 + 35 = 2 + y$

25. $\frac{2}{5} = x - \frac{9}{20}$

26. $n - 6 = 99 + 3$

Solve these algebraic equations for the unknown variable using multiplication/division properties of equality.

27. $6x = 72$

28. $48 = \frac{3}{2}x$

29. $10y = 4y + 30$

30. $60 - x = 4x$

31. $\frac{1}{3}x = 99$

32. $4x = \frac{1}{4}$

33. $5.6\,y = 28$

34. $9.4\,a = 37.6$

35. $66.5 = 13.3\,b$

36. $\frac{h}{15} = 11.5$

Middle School Math Quest

Algebraic Tug of War

Help balance each Tug of War competition. Find the value of the missing colors.

- The values on either side of the rope must equal each other.
- People that are the same color all pull with the same strength.

● = ? + ? =

4 = 2 + 2

If Purple pulls with a strength of 4, how much force do the Blues need to use? Since they're the same color they must be pulling with the same force, so they are pulling with a strength of 2.

Find the strength of each different color stick figure.

1.

12

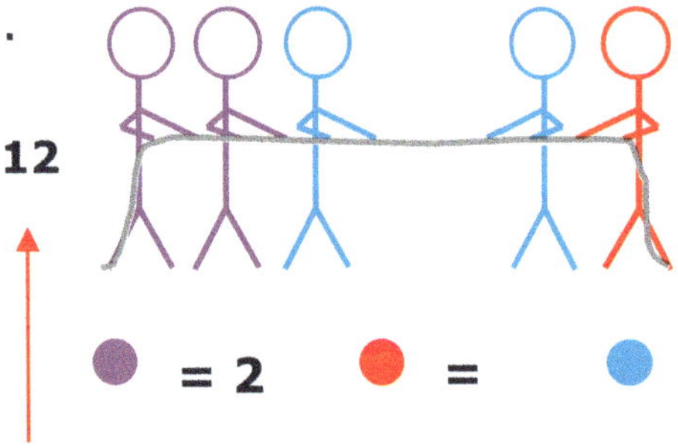

● = 2 ● = ● =

Numbers next to groups of people show how hard that group is pulling.

2.

● = 1 ● = ● = 9

3. There are two groups on each side of the ropes intersection. The left and right sides must pull with equal strength. However, each team on the same side can pull with a different strength than the other team on that side.

15

18

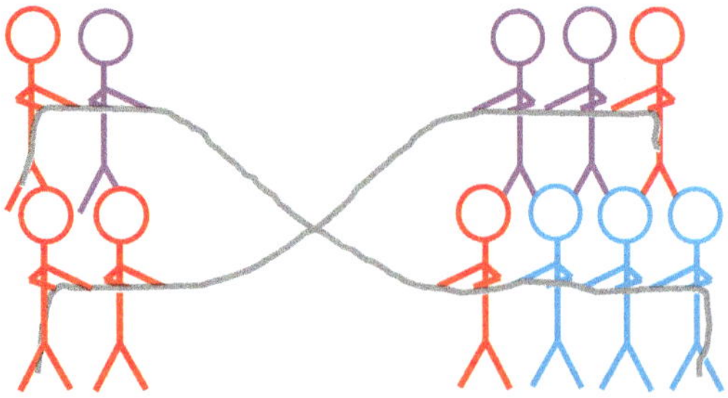

● = ● = 9 ● =

4. First, determine the color strengths needed to balance the left and right sides. Then, **using the fewest number of people possible,** determine how many people of each color would you need to add to the left side for them to have 17 more strength than the right side.

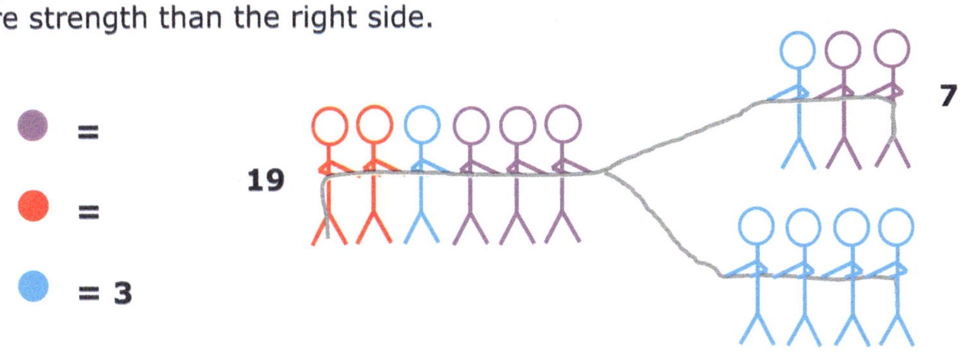

5. **Purple** is pulling with each arm to balance the weight applied by the left and right sides. How hard are the **Pinks** pulling? How hard is **Purple** pulling on the right side?

6. The two groups on the left side need to balance the three right-hand groups. Solve for the remaining colors.

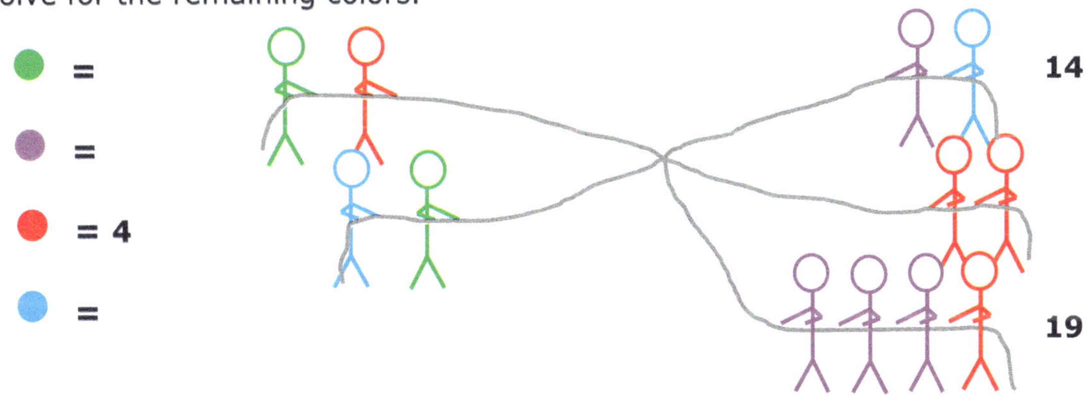

With this multi-part question, whatever you solve for earlier in the problem remains the same for the following parts.

7a. **Purple** is pulling with a strength of **40.** Solve for **Red**.

7b. Some **Reds** and **Greens** show up on the right side and start pulling to help **Purple**. How much of **Purple's** force of 40 (above) are they able to take over?

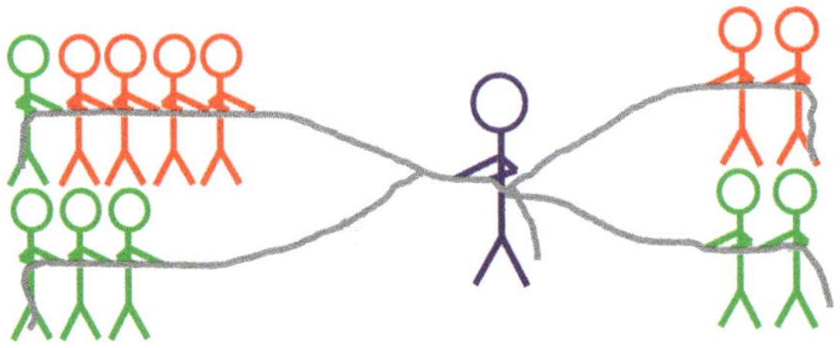

7c. **Purple** leaves to walk his dog and some of the **Reds** and **Greens** have to go to school, so a couple **Blues** come over to join the fun. Solve for **Blue's** strength.

If the right side wanted to pull at least three times harder, how many more **Blues** would they need to help them?

8a. Two teams are pulling back a catapult so they can launch their friends into a castle. They need to pull with a strength of 80 total and **the two teams are pulling with equal strength**. Find how hard each color is pulling.

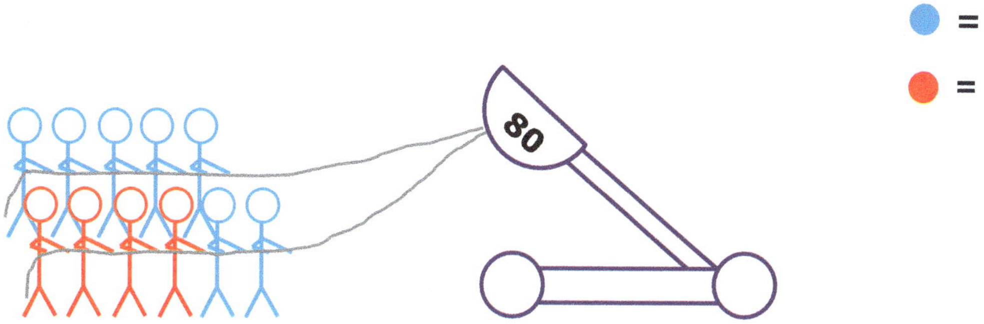

8b. Each person weighs as many pounds as they can pull. The catapult can launch up to 80 pounds. Find the maximum number of people that can fit in the catapult. Are there different combinations of colors that give you the same maximum number of people?

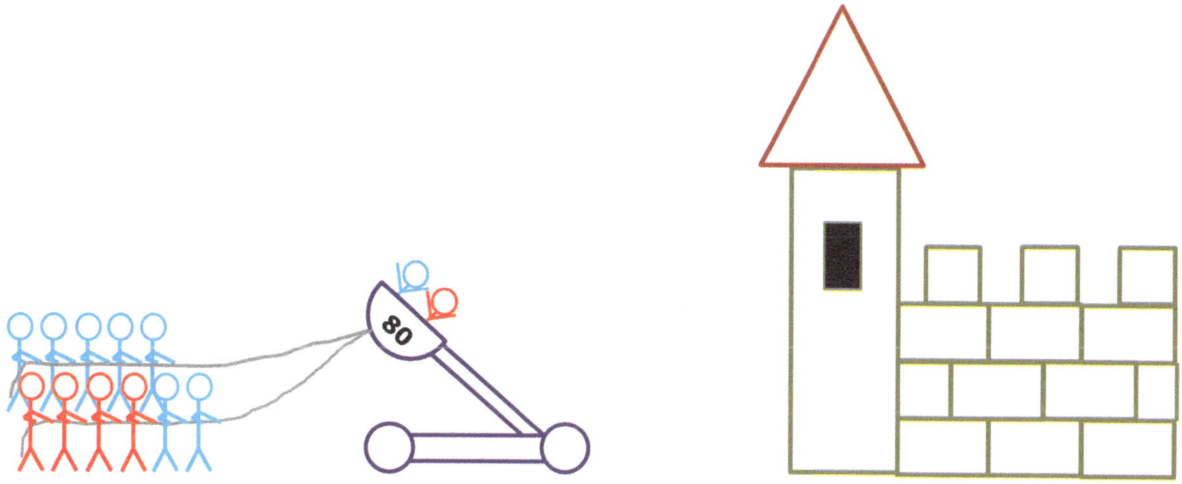

9. **Tim** is pulling up vine people. Each vine weighs exactly 20 pounds. Solve for the remaining colors.

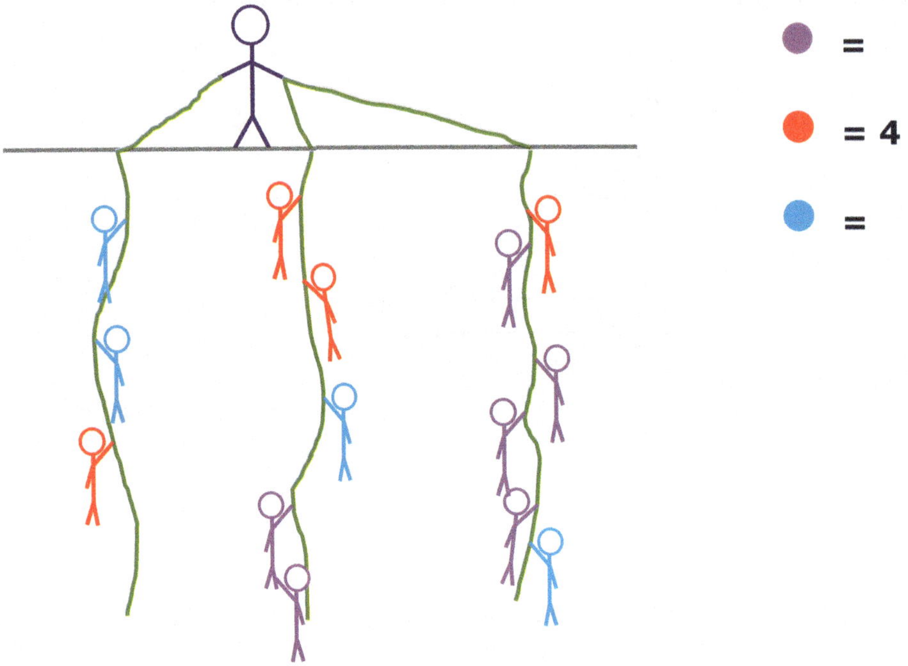

10. **Angela** is also lifting vines that can weigh at most 20 pounds. Solve for each color using the left and middle vine information. Then, see if **Angela** can pull the right vine up on her own. If she can't, find out how many **Angelas** you would need to pull up the right-hand vine.

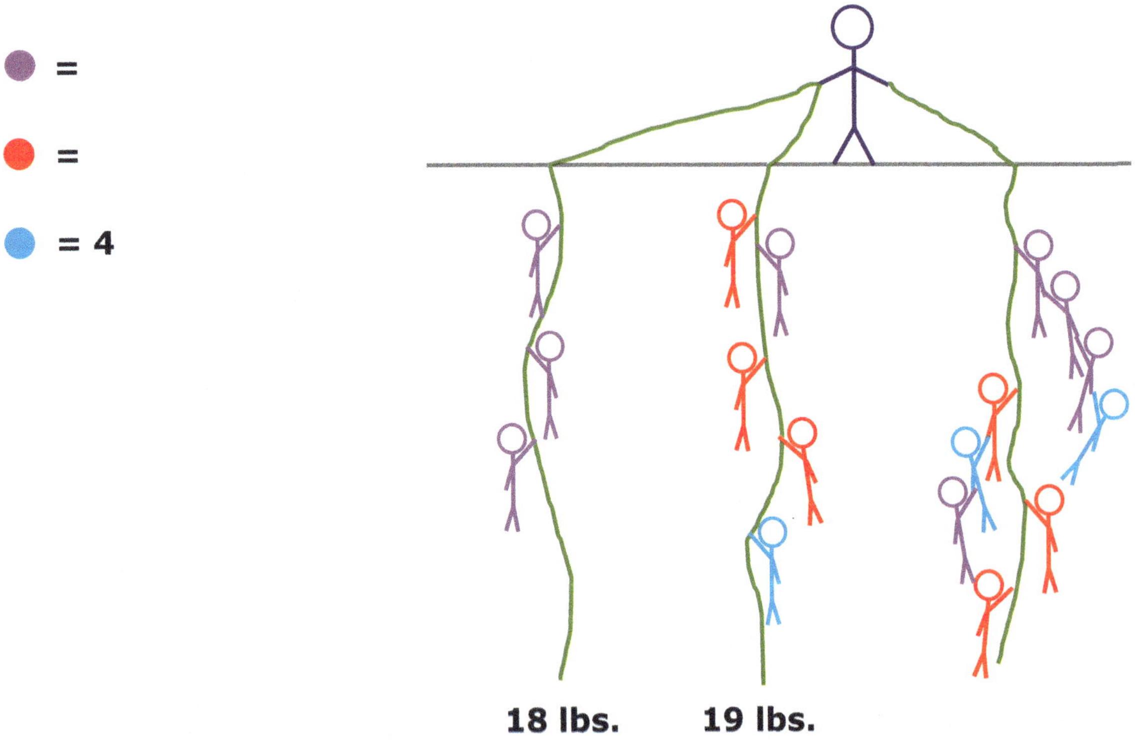

Probability Palooza

One of the most important skills needed to calculate probabilities is the ability to count. Since you learned that a long time ago, you won't have any problems mastering probability.

Probability Fraction Definition

Example: Exploring Dice Rolls

Let's roll one die.

List of all outcomes: 1, 2, 3, 4, 5, 6 **Total Number of Outcomes**: 6

How likely is it to roll an even number?

List of **Favorable Outcomes**: 2, 4, 6

We are likely to roll an even number 3 out of 6 times

$$\textbf{Probability}\ of\ rolling\ an\ even\ number = \frac{\#\ of\ Favorable\ Outcomes}{\#\ of\ Total\ Outcomes} = \frac{3}{6} = \frac{1}{2}$$

Note: List outcomes or give a whole number if questions ask you for **total**, or **favorable**, outcomes. Use a fraction to give the probability of a **favorable** outcome.

Use the listing and counting strategy shown in the box above to calculate these probabilities.

1. What is the probability of rolling greater than a 1 with one die?

2. What is the probability of rolling less than 3 with one die?

3. What is the probability of flipping two coins and both landing on Heads?

4. What is the probability of rolling two dice where their sum is greater than 10?

Middle School Math Quest

Listing Outcomes with Tables and Tree Diagrams

Example: Show all possible outfits you can make if you have a Red and Blue Shirt along with Purple, Yellow and Green Pants.

Tree Diagram Approach

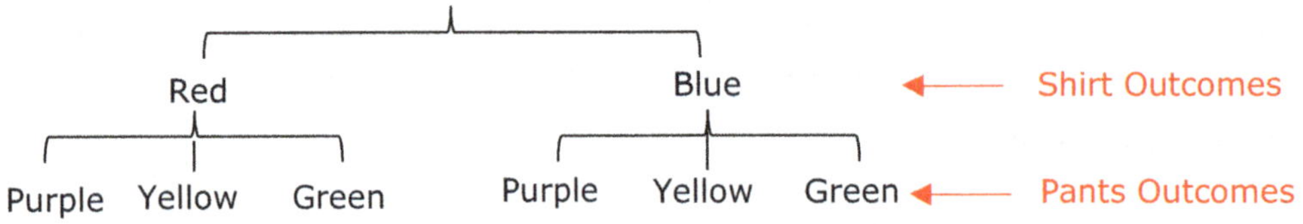

Our tree diagram shows us 6 possible outcomes: (Red Shirt, Purple Pants), (Red Shirt, Yellow Pants), (Red Shirt, Green Pants), (Blue Shirt, Purple Pants), (Blue Shirt, Yellow Pants), (Blue Shirt, Green Pants)

Table Approach

Shirts \ Pants	Purple	Yellow	Green
Red	(R,P)	(R,Y)	(R,G)
Blue	(B,P)	(B,Y)	(B,G)

Use the tree diagram or table approach to list all possible outcomes and calculate probabilities.

5. How many ways are there to roll two dice? List all possible ways (Hint: The table format may be most helpful.) Check the answer key for this question before you move on to the rest of the questions in this section.

6. What is the probability of rolling two dice, both greater than 4?

7. How many outcomes are there with a sum greater than 7 when you roll two dice?

8. What is the probability of rolling two dice with a sum greater than 7?

9. What is the probability of rolling two dice with a sum greater than 1?

Answer these questions using your best counting and outcome listing strategies.

10. A bag contains balls numbered from 28 through 45. What's the probability of pulling a prime number?

11. On Track Day, there is a long race on the field. To help the runners know how far they have to go, flags are numbered every 5 yards, starting at 50 yards out (50, 55, 60,...). It's a windy day and the wind is equally likely to blow over each flag. If the race ends at 230 yards, what's the probability the wind knocks over a flag ending with a zero?

Multiplication Principle

Using the tree or table method, we listed all possible outcomes. If you only need the total number of outcomes, you can instead use the **Multiplication Principle**.

Multiplication Principle: If you have a ways of choosing one outcome and b ways of choosing another outcome, your total number of outcomes will be $a \times b$.

Total Outcomes = # of ways to choose 1st choice x # of ways to choose 2nd choice

Using our previous example, we had 2 shirts and 3 pants. We can choose shirts 2 ways and pants 3 ways so our total outcomes are $2 \times 3 = 6$.

Answer these questions using the multiplication principle:

12. If you roll one die, spin a spinner with 5 letters on it and flip a coin, how many outcomes are there?

13. You need to create a 5-digit security code for your smart phone. How many codes can you make if your code can't start with a zero and you must use a different digit for each of your 5 numbers?

Middle School Math Quest

Geometry Concepts

Area of a Triangle

$A = \frac{base \times height}{2} = \frac{bxh}{2}$

Use right angle height, **not** slant height.

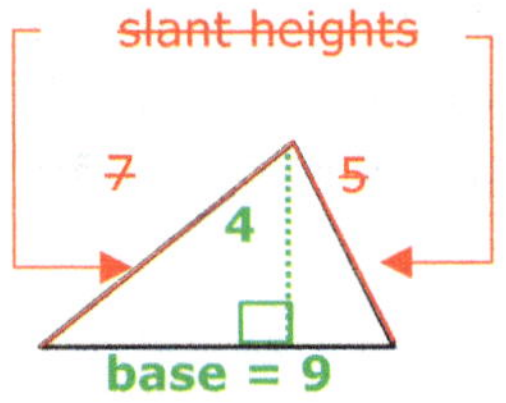

Area $= \frac{9 \times 4}{2} = 18$

Triangle Inequality

The sum of the lengths of two sides of a triangle is greater than the length of the third side.

Sum of Triangle Angles

The sum of all the angles in a triangle is 180°.

Triangle Classifications

By Angle:

Acute (All angles < 90°)
Right (one 90° angle)
Obtuse (one angle > 90°)

By Side Length:

Scalene
(All sides different)
Isosceles
(2 sides the same)
Equilateral
(3 sides the same)

Quadrilateral Classes

Quadrilaterals have 4 sides and 4 angles that sum to 360°.

Parallelograms have opposite parallel sides and opposite angles of the same measure.

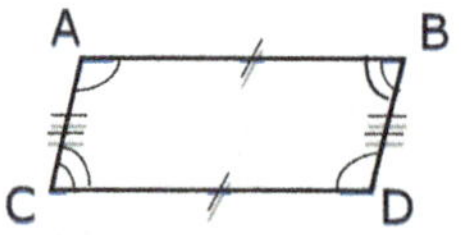

$m\angle A = m\angle D;\ m\angle B = m\angle C$

Rhombuses are **parallelograms** with all sides equal length.

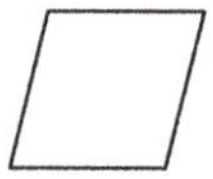

Rectangles are **parallelograms** with 4 right angles.

Squares are **rectangles** with all sides equal length.

Trapezoids have one pair of opposite parallel sides. Angles between the parallel sides sum to 180°.

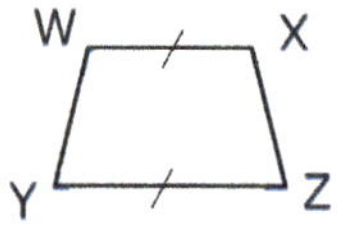

$m\angle W + m\angle Y = 180°;\ m\angle X + m\angle Z = 180°$

Angles on a Line

Adjacent angles on a line sum to 180°.

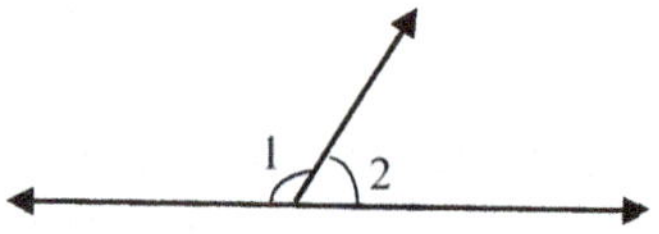

$m\angle 1 + m\angle 2 = 180°$

Angles at a Point

Angles at a point sum to 360°.

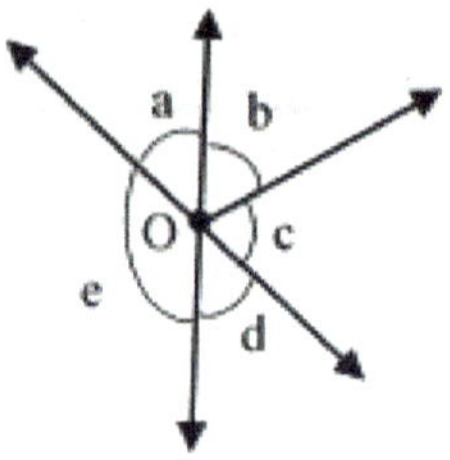

$m\angle a + m\angle b + m\angle c + m\angle d + m\angle e = 360°$

Vertical Angles

Opposite angles formed by 2 lines are vertical angles and have the same measure.

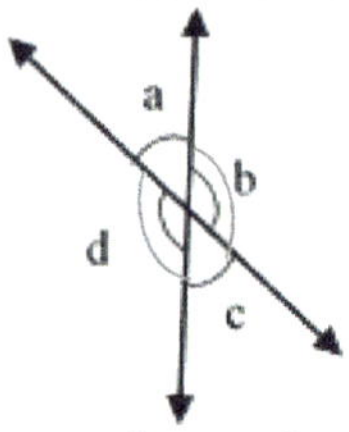

$m\angle a = m\angle c;\ m\angle b = m\angle d$

Volume of Rectangular Prisms

Volume tells us how much a container can hold. For cubes, the formula is:

$V = length \times width \times height$

or

$V = area\ of\ base \times height$

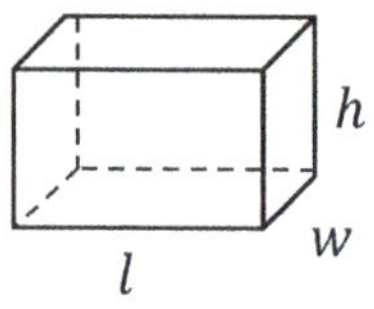

Use the previous definitions page to answer these questions.

Find the unknown angle measures.

1. Using vertical angles

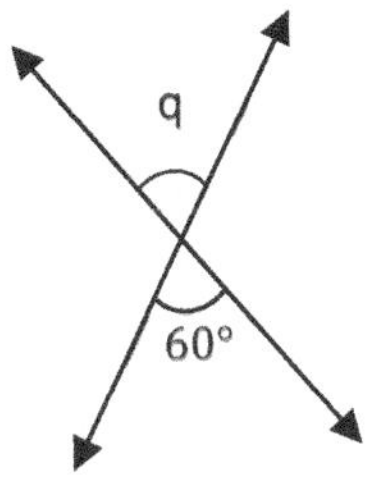

q = __________

2. Using angles on a line and at a point; vertical angles

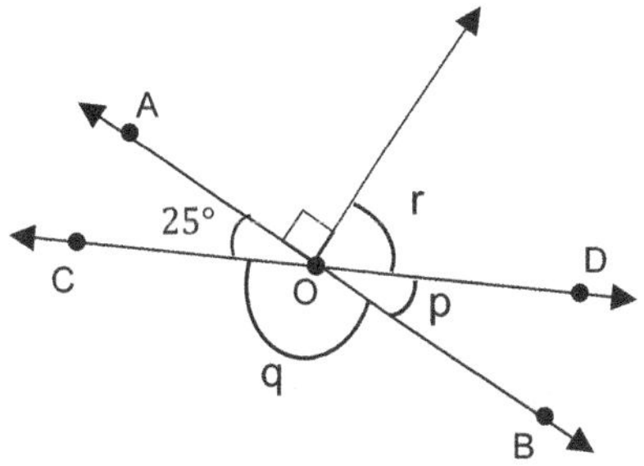

p = __________

q = __________

r = __________

Find the unknown angle measurement and state whether the triangle is scalene, isosceles or equilateral.

3.

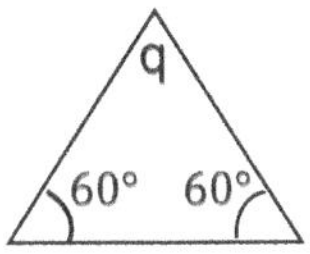

4.

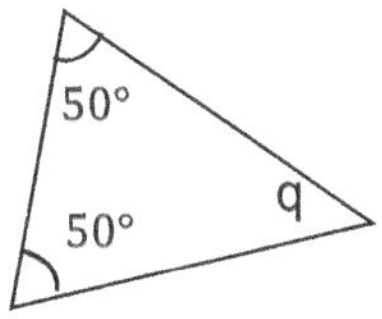

5.

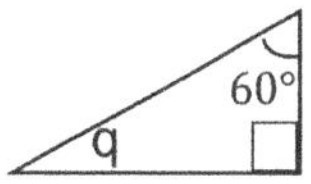

q = __________ q = __________ q = __________

Triangle: __________ Triangle: __________ Triangle: __________

Do you feel confident with angle measurements now? If not, you can create some of your own problems using a protractor and ruler – Go crazy with what you create and notice patterns as you explore.

Middle School Math Quest

Perimeter and Area Concepts

Rectangle Area

$A = l \times w$

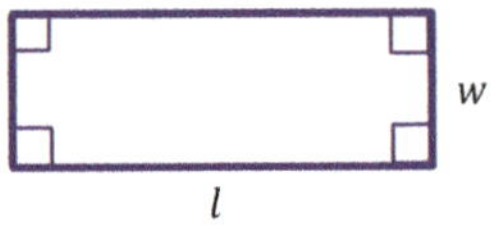

Parallelogram Area

$A = b \times h$

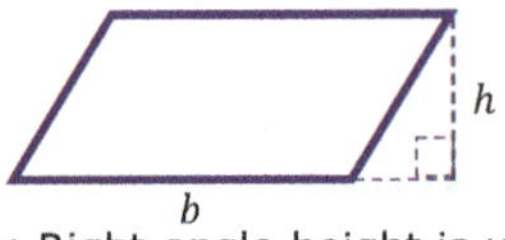

Note: Right-angle height is used for this calculation.

Trapezoid Area

$A = \frac{(b_1 + b_2) \times h}{2}$

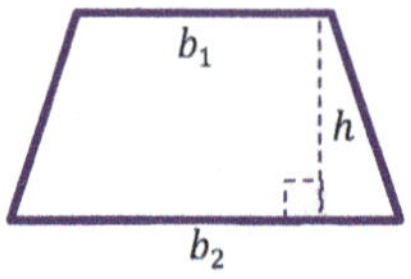

Prism Definition

A **prism** is a solid geometric shape with two parallel and congruent (same shape) bases. Its sides are all parallelograms.

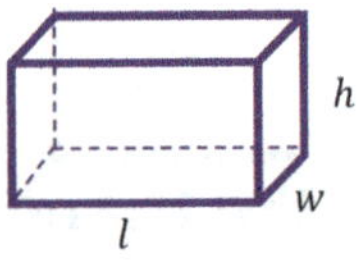

Prism Surface Area

$SA = Sum\ of\ each\ Face's\ area$

Ex: If a rectangular prism has length = 5, width = 4 and height =3, then we have:

SA = 2(5x4 + 5x3 + 4x3) = **94 square units**

Polygon Definition

A **polygon** is a two-dimensional, closed shape made of only straight-line segments (curved shapes are NOT polygons.)

Pyramid Definition

A **pyramid** is a solid geometric shape with one polygonal base and polygonal sides. The sides of the shape meet at the apex (a point at the top of the pyramid.)

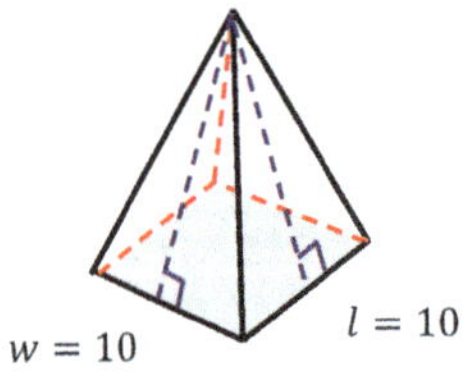

Pyramid Surface Area

$SA = Sum\ of\ each\ Face's\ area$

Ex: Using the pyramid at the left, if the height of each triangular face is 15, then we have:

$\mathbf{SA} = 10 \times 10 + \left(4 \times \frac{1}{2} \times 10 \times 15\right)$

Square Area + Triangle Areas

= **400 square units**

Polygon Perimeter

The perimeter of any **polygon** is the sum of all the edge lengths that form the shape.

With the information on this page and on page 40, you have everything you need to answer the following questions. Let's practice some more geometry!

6. The area of a rectangular dining room is 96 square meters. It is 8 meters wide. How long is it?

7. A square piece of gold has an area of 64 square millimeters. How long is each side of the gold piece?

8. A beach volleyball court is 8 meters wide and 17 meters long and uses a rope to mark the boundary sides and ends. The rope costs $1.93 per meter. How much would it cost to buy rope to make the boundary lines for this court?

9. The average length of a trapezoid's bases is 8 cm and its height is 6 cm. What is the area of this trapezoid?

Find the area of the following figures:

10.

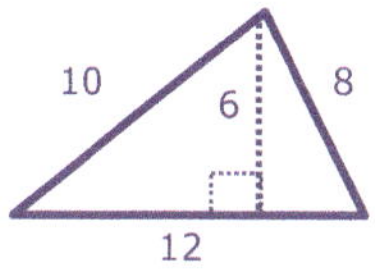

11.

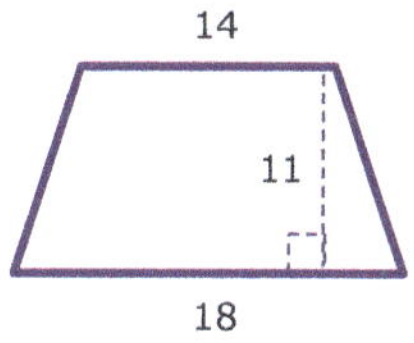

12.

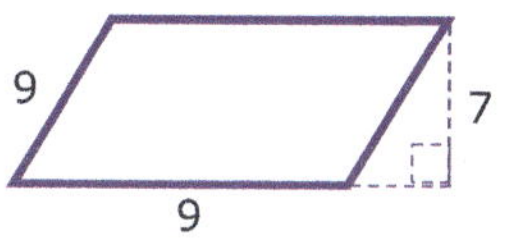

13. What is the surface area of a rectangular prism with dimensions 4 x 6 x 8?

14. What is the volume of the same rectangular prism from question 13?

15. Molly is creating a scale model of one of the square Egyptian pyramids. She plans to cover each face, including the base, with tan fabric resembling limestone. If each triangular face of her model has an area of 32 square inches and her square base has a side measuring 6 inches, how much fabric will she need?

Middle School Math Quest

Radical Ratios

Ratios are a fun topic to learn; you'll use ratios the rest of your life. Ratios are used to create maps, describe rates (miles per hour) and convert recipes from 4 servings to 40 servings.

Ratios Overview

CONCEPTS	EXAMPLES
Ratios compare two or more categories.	We have 4 chocolate and 3 fruit candies
Ratios can be represented using words.	4 chocolate to 3 fruit candies or 3 fruit to 4 chocolate candies
Ratios can be represented using fractions.	Chocolate to Fruit Ratio $= \frac{4}{3}$ Fruit to Chocolate Ratio $= \frac{3}{4}$
Ratios can be written in colon form.	Chocolate:Fruit Ratio = 4:3
For a given situation, we can write 6 ratios based on that situation (Note: These ratios are NOT equivalent but they do describe the same situation).	Our 6 ratios are: $\frac{4 Choc}{3 Fruit}, \frac{3 Fruit}{4 Choc}, \frac{4 Choc}{7 Total}, \frac{7 Total}{4 Choc}, \frac{3 Fruit}{7 Total}, \frac{7 Total}{3 Fruit}$
We can use bar modeling to represent ratios.	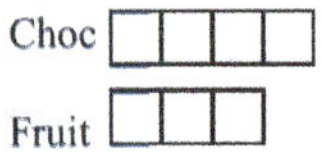

Practice using ratios to describe the relationships between these shapes.

1. What is the ratio of squares to circles? Write your ratio in word form.
2. What is the ratio of triangles to squares? Write your ratio in fraction form.
3. What is the ratio of triangles to circles? Write your ratio in colon form.
4. What is the ratio of circles to the total number of shapes?

Additional Ratio Representations

Example: We have a tattoo kit of emojis

Tattoo Kit ☺☺☺☹☹

The ratio of smiles to frowns is:

3:2 *or* $\frac{3}{2}$ *or* 3 smiles to 2 frowns *or*

What if we had two of these tattoo kits? What would be the ratio representing smiles to frowns?

Tattoo Kit 1 ☺☺☺☹☹

Tattoo Kit 2 ☺☺☺☹☹

We could write our ratio as $\frac{6}{4}$ because we have 6 smile emojis and 4 frown emojis. Since 6 and 4 have a common factor, though, we could STILL represent the ratio as $\frac{3}{2}$ because for every 3 smile emojis we have 2 frown emojis.

Equivalent ratios are the same as equivalent fractions.

Ratio of $\frac{6}{4}$ = Ratio of $\frac{3}{2}$

Example: Sue has a total of 20 apples and oranges. If the ratio of apples to oranges is 3:1, how much of each fruit does she have?

We can use bar modeling to represent the ratio and solve:

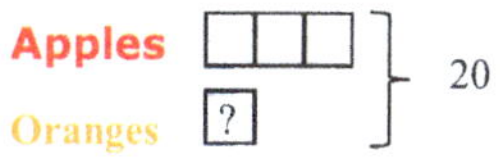

$$20 \text{ total} \div 4 \text{ parts} = 5 \text{ fruits per part}$$

Answer: There are 5 oranges and 15 apples.

We can also use equivalent fractions/ratios to solve this example.

Let's choose a ratio that includes total fruits since we are given that information.

$$3 \times 5 = 15 \text{ apples}$$

$$\frac{3_{\text{Apples}}}{4_{\text{Total}}} = \frac{?_{\text{Apples}}}{20_{\text{Total}}}$$

$$4 \times 5 = 20$$

We have 15 apples and 5 oranges.

Practice your new ratio skills with these exercises.

5. In a 6th grade classroom of 30 students, the ratio of girls to boys is 3:2. How many girls are there, and how many boys?

6. Dory is pet-sitting 12 dogs and cats. A third of the animals are cats. What is the ratio of cats to dogs at her house?

7. John owns 10 shirts and 6 pairs of pants. He goes shopping and buys 3 more pairs of pants. How many more shirts does he need to buy for the original ratio to remain the same? What is that ratio?

8. Catelyn has 25 red marbles and 10 blue marbles. What is the ratio of blue to total marbles? Write your answer in simplest form.

Middle School Math Quest

Rates

Rates are ratios that compare quantities with different units of measurement. **Unit rates** use equivalent fractions to represent a quantity in terms of **one unit** of the other quantity. Here are some examples:

Rates	Transformation	Unit Rates
30 miles per gallon	No change needed because we already represent the ratio per 1 gallon	30 miles per gallon
3 fidget spinners for $10	We'll divide $10 by 3 so we know how much each spinner costs	$3.33 per fidget spinner
$40 per 5 weeks	Let's find out how much per week (divide $40 by 5)	$8 per week

Try these ratio challenge problems. Think carefully and use all the tools you've learned.

9. A class tallied what color of shirt each student was wearing on Monday. If the probability that a student wore a blue shirt is 6 out of 10, what is the ratio of blue shirts to other colored shirts in the class?

10. Voters were choosing between Ms. White and Mr. Martinez as mayor. If the probability a voter voted for Ms. White is 7 out of 10 and there was a total of 28 voters voting for Ms. White, how many voters were there in all?

11. If you make $34 for 2 hours of work, how much money will you make after working 5 hours?

12. Greg can swim 24 laps in 6 minutes and Bob can swim 2 laps in 40 seconds. Who swims faster?

13. For every $2 that Ben spends, Clara spends $3. If Clara spends $60 more than Ben, how much do they spend altogether?

14. Which car gets better gas mileage?

Car	**A**	**B**
Distance (miles)	675	784
Gallons Used	27	28

15. Which is the better buy, Box A or B?

Cupcakes	**Box A**	**Box B**
Cost (dollars)	675	784
Number of Cupcakes	27	28

Middle School Math Quest

Percents Plus

Like ratios, you'll use percentages frequently. This symbol, %, is the percent sign and is read "percent". Stores advertise, "20% off through Sunday" and news stories talk about a "50% increase in productivity over the past 10 years." We're going to teach you how to calculate, and interpret, percentages.

Percents Concepts

Definition: A **percent** represents "How many parts out of 100."

50% means 50 parts out of 100 OR $\frac{50}{100}$ OR $\frac{1}{2}$

50% of these shapes are squares

Question: Why do we use percents instead of fractions?

Answer: Sometimes it's hard to compare fractions because they have different denominators. By converting fractions to percents, you know exactly which percent is bigger because they have the same denominator, 100.

Example: Which fraction is bigger, $\frac{68}{75}$ or $\frac{84}{93}$?

It would take a lot of work to calculate a common denominator for these fractions. Instead, if we know how to calculate percentages, we would see:

$\frac{68}{75} = 90.67\%$ and $\frac{84}{93} = 90.32\%$ → $\frac{68}{75}$ **is larger than** $\frac{84}{93}$

Percent and Fraction Relationships

Example: What fraction of the square is shaded? Can you write the shaded part as a percent?

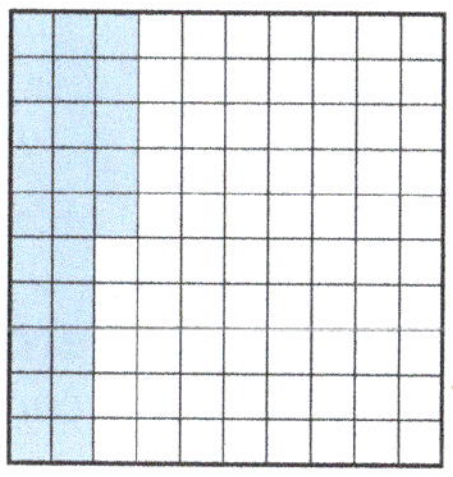

Answer:

$\frac{25}{100}$ of the square is shaded → 25 out of 100 parts are shaded which is **25%**.

We could also say $\frac{1}{4}$ of the square is shaded (simplest form of our fraction).

Write the fraction and percent representations for the shaded parts of these squares. Use simplest form for your fractions.

Q1

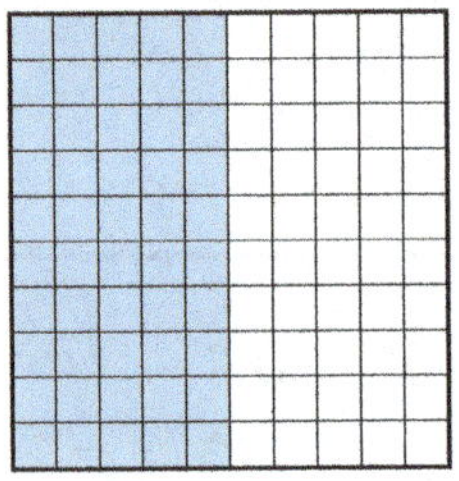

Q2

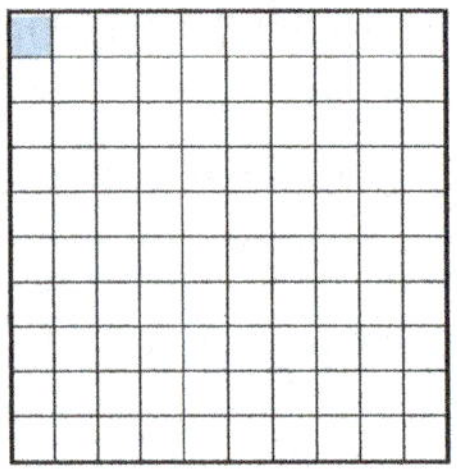

Q3

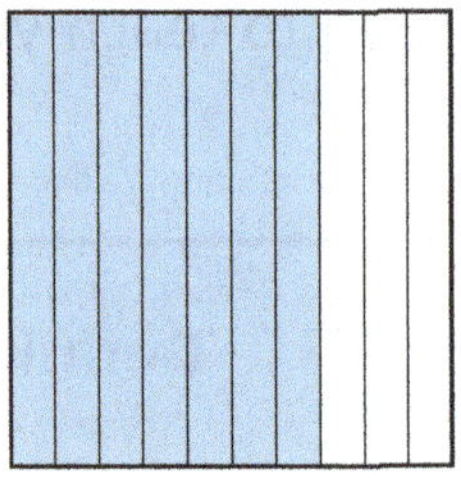

Hint: How many hundredths are in tenths?

Now that you've mastered writing percents, we're going to show you how to convert between percent and decimal representations of numbers.

Writing Decimals as Percents

To write a decimal as a percent, we can first think about converting decimals to fractions.

Example: Convert 0.05 to a percent.

$0.05 = \frac{5}{100}$ and $\frac{5}{100} = 5\%$ → $\mathbf{0.05 = 5\%}$

Example: Convert 0.2 to a percent.

$0.2 = \frac{2}{10} = \frac{20}{100}$ and $\frac{20}{100} = 20\%$ → $\mathbf{0.2 = 20\%}$

Question: What patterns can we use to convert between decimals and percents?

From our examples above, we saw: $0.05 = 5\%$
$0.20 = 20\%$

Decimal and Percent Conversion Rules

Decimals to Percents: We can convert decimals to percents by moving the decimal 2 places to the right and adding the percent sign.

Example: 0.005 = 0.5%

Move decimal point 2 places to the right.

Percents to Decimals: We can convert percents to decimals by moving the decimal 2 places to the left and removing the percent sign.

Example: 125% = 1.25

Move decimal point 2 places to the left.

Middle School Math Quest

Use your new tools to fill in this conversion table. Simplify all fractions.

Fraction, Decimal, and Percent Conversions

FRACTION	DECIMAL	PERCENTAGE
$\frac{5}{8}$		
	0.1	
		$66.\overline{66}\%$
	0.40	
$\frac{1}{5}$		
	0.05	
	0.24	
		8%
1		
		75%
		30%
$2\frac{3}{4}$		
	1.25	
		$33.\overline{33}\%$

FRACTION	DECIMAL	PERCENTAGE
$\frac{17}{20}$		
	0.99	
		80%
$\frac{1}{4}$		
	0.125	
$1\frac{5}{6}$		
$\frac{19}{20}$		
	0.45	
$\frac{1}{6}$		
		55%
		$77.\overline{77}\%$
	0.18	
	0.01	
		90%

We've converted between fractions, decimals and percentages. Now, we'll teach you the bar modeling skills you need to solve percentage problems.

Percents with Bar Modeling and Decimals

Bar modeling can help you visualize percent problems. Bar models can help you think about more difficult percent problems.

Ex 1: What is 40% of 50?

Hmmm.... $40\% = \frac{2}{5}$

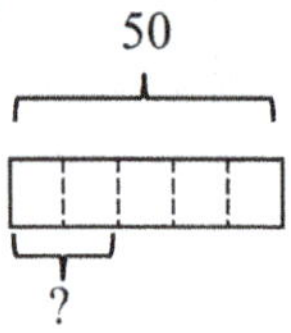

$\frac{2}{5}$ of $50 = \frac{2}{5} \times 50 = 20$

40% of 50 is 20

Ex 2: 15 is what percent of 60?

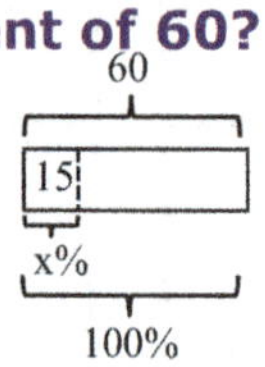

Set up part-to-whole ratios

$\frac{15}{60} = \frac{x}{100}$ or $\frac{1}{4} = \frac{x}{100}$

We got this by simplifying the left-hand ratio.

Use equivalent fractions to get:

$\frac{1}{4} = \frac{25}{100}$

15 is 25% of 60

Ex 3: 70 is 50% of what number?

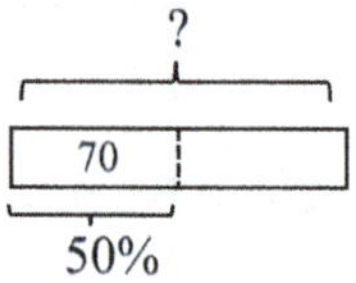

We could set up part-to-whole ratios BUT we notice we can multiply 70 by 2 to get our answer.

70 is 50% of 140

Percents with Decimals

Converting a percent to a decimal can help with quick mental math calculations for basic percentages.

Cool Strategy for 10%: To find 10% of a number, just move the decimal place in the number one place to the left.

Why does this work? Because 10% = 0.10. And, from your Decimal Discovery work, you know multiplying by 0.1 moves the decimal point one place to the left.

Ex 4: 10% of 90 = 9

Ex 5: 20% of 90 = 2 x 10% of 90 = 2 x 9 = 18

Cool Strategy for 1%: To find 1% of a number, just move the decimal place in the number two places to the left.

Why does this work? Because 1% = 0.01. And, from your Decimal Discovery work, you know multiplying by 0.01 moves the decimal point two places to the left.

Ex 6: 1% of 90 = 0.9

Ex 7: 2% of 90 = 2 x 1% of 90 = 2 x 0.9 = 1.8

Use your new percentage skills to solve these problems.

1. 75% of 120

2. 80% of 60

3. 90% of 50

4. 60 is 40% of what number?

5. 150% of 40

6. 5% of 120

7. 200% of 30

8. 10 is 25% of what number?

9. 30 is what percent of 50?

10. 120 is what percent of 90?

11. 80 is what percent of 200?

12. 25 is what percent of 5?

13. If 40% of the boys and 30% of the girls in a class like baseball, how many students like baseball if there 20 boys and 30 girls in the class?

Domino Ratios

Now that you've learned about ratios let's apply them to some puzzles. In this example we're given three tiles and we need to fill in the given grid.

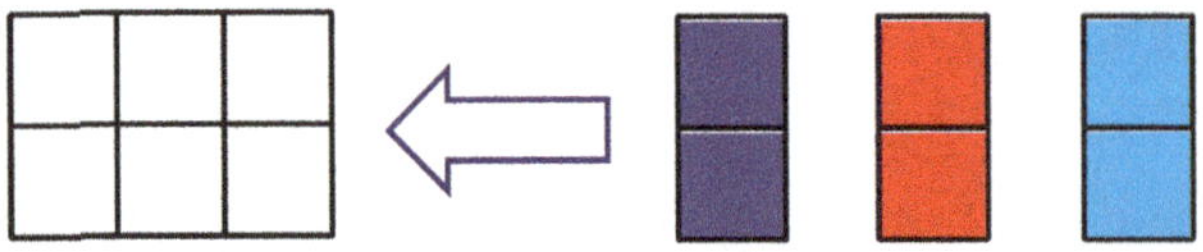

- What if we wanted to create a grid with a 1:1:1 color ratio?

- Or a 1:2 ratio of red to blue?

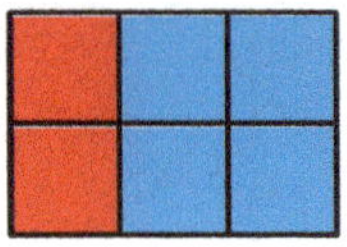

Now use your puzzle solving skills to solve the rest of these grids.

1. Using these tiles, make a 2:5:7 blue:yellow:red ratio. In addition, make sure all blue squares touch each other. You can use as many of each tile as you want, just make sure all the white squares get filled in. You cannot fill in the grey squares – they remain blank for all of these challenges.

2. Using the tiles on the right, fill in the top half so it has a 1:2:3 ratio of red, yellow, and blue. Then, can you fill in the bottom half to make a 3:2:1 ratio of red, yellow, and blue using the same tiles?

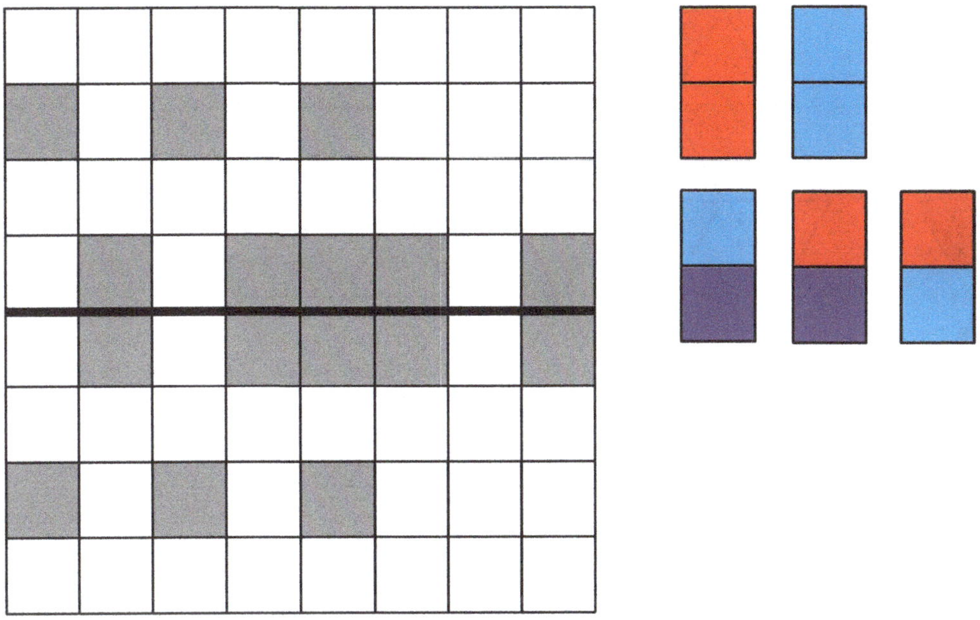

3. Make a 1:2:6 ratio of blue, yellow, and red using these larger tiles.

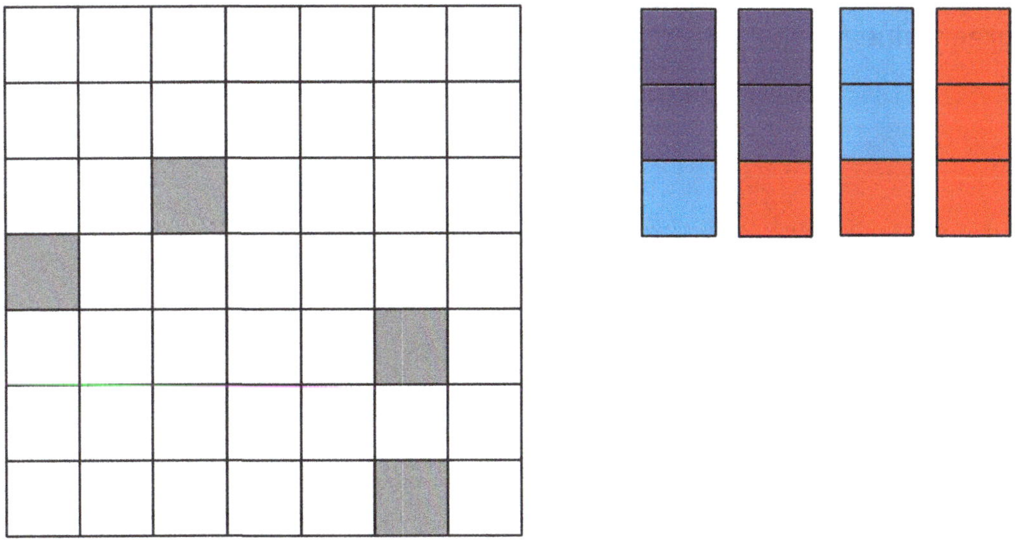

4. Now it's time to try a problem with new shapes!

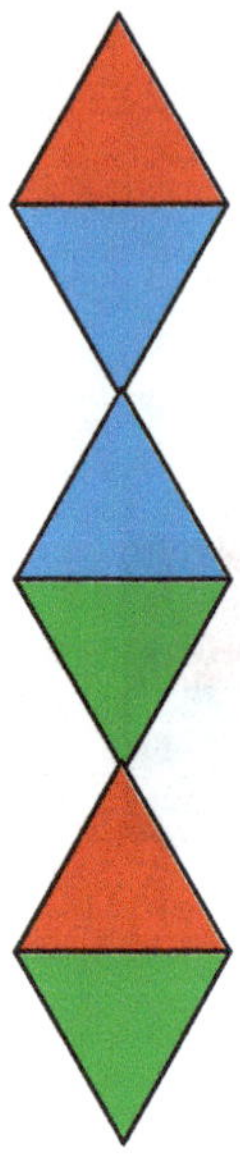

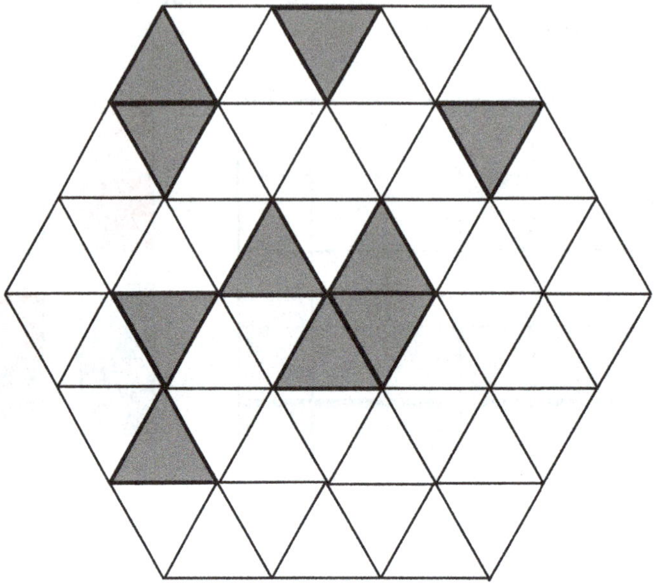

What is the maximum number of spaces you can fill red using the tiles shown above? Write the ratio of all three colors with the tiles that you used to get the maximum number of red spaces. Is there a way to create a more simplified ratio of red:green:blue for the pattern you used?

Intro to Exponents

You may have seen some exponents ideas already at school. We'll review basic concepts and introduce you to some more cool tricks.

Remember....

→ **Multiplication is repeated addition**

3×2 is telling you to "Take three 2s and add them together."

$3 \times 2 = 2 + 2 + 2 = 6$

→ **Exponents are repeated multiplication**

2^3 is telling you to "Take three 2s and multiply them together."

$2^3 = 2 \times 2 \times 2 = 8$

Exponents Overview

- A **power** is the answer to a repeated multiplication problem.
- The **base** of the power is the number you repeatedly multiply.
- The **exponent** of the power tells you how many times to repeat your multiplication.

Power, Exponent, Base

$2^4 = 2 \times 2 \times 2 \times 2 = 16$

Repeated multiplication of **2**, **four times**

- A **perfect square** is the answer to a power problem with an exponent of 2.

$1^2 = \mathbf{1}$
$2^2 = \mathbf{4}$
$3^2 = \mathbf{9}$

1, 4, 9, 16, 25, ... are perfect squares.

Ex 1: $6^3 = 6 \times 6 \times 6 = 216$

Ex 2: $4^4 = 4 \times 4 \times 4 \times 4 = 256$

Ex 3: 36 is a perfect square because $6^2 = 36$.

Ex 4: 94 is not a perfect square because the only factors of 94 are 2 and 47.

Use the overview and examples from the previous page to help you find the answers to these problems:

1. 4^2

2. 5^3

3. 3^4

4. 10^3

5. 11^3

6. 1^4

7. 2^{10}

8. 3^3

9. 2^4

10. 9^3

11. List the first 20 perfect squares.

12. A perfect cube is the answer to a repeated multiplication problem with 3 as the exponent. 8 is a perfect cube because it's the answer to the exponent problem 2^3. List the first 10 perfect cubes. Are there any perfect cubes and perfect squares from this question and question 11 that are the same?

13. What is the base in the exponent expression, 10^4?

Prime Factorization

You already know how to find all the factors of a number (do you remember rainbow, or mountain, factoring from 4th grade?). Now, we're going to find the prime factors that multiply together to make a number. You'll start with factor trees.

Prime Factorization Steps

STEP 1: Let's make a factor tree for 40. Think of two numbers that multiply to 40 ($5 \times 8 = 40$):

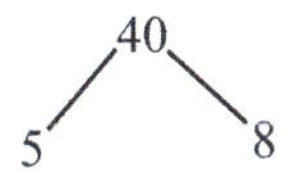

STEP 2: If either of the factors we put in the factor tree are prime numbers, we **circle** them and continue making new branches from any composite numbers by thinking of more multiplication facts. We expanded the 8 branch by using $4 \times 2 = 8$.

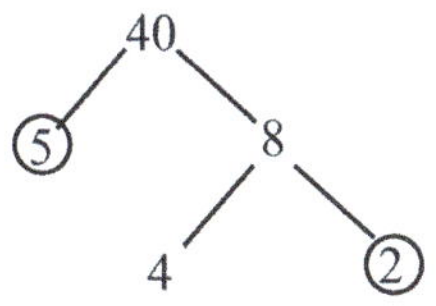

STEP 3: Keep repeating this process until all the branches end in circled primes.

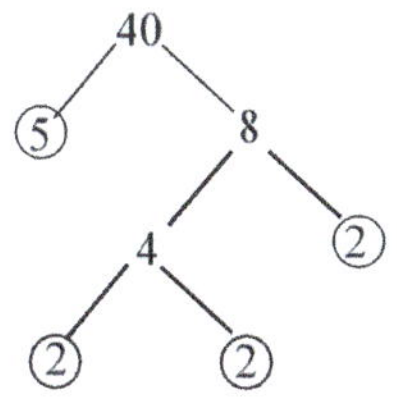

STEP 4: Write the prime factorization of the number by showing the prime numbers (the numbers you circled) that multiply to your original number.

The prime factorization of 40 is $2 \times 2 \times 2 \times 5$.

The exponent form of this would be $2^3 \times 5$.

A Few More Examples....

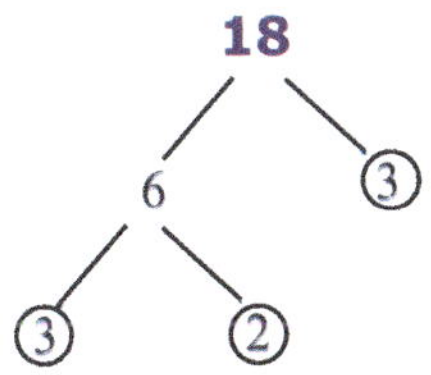

The prime factorization of 18 is 2x3x3 or 2×3^2.

The prime factorization of a prime number is just itself, so the prime factorization of 7 is 7.

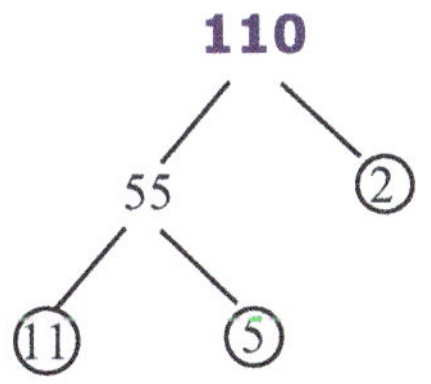

The prime factorization of 110 is 2x5x11.

Now it's your turn. Find the prime factorization of these numbers:

1. 12
2. 84
3. 64
4. 60
5. 37
6. 81
7. 26
8. 1,000
9. 256
10. 36

You just mastered a middle school concept!

Middle School Math Quest

Prime Factorization, GCF and LCM

In elementary school, you learned how to find the GCF (Greatest Common Factor) and LCM (Least Common Multiple) of numbers. Now, we'll show you how to find these using prime factorization techniques. Remember, prime factorization gives us the recipe for each number.

Finding GCF and LCM using Prime Factorization

Step 0: Find the Prime Factorization of each number

$$30 = 2 \times 3 \times 5$$
$$18 = 2 \times 3 \times 3$$

Step 1: Map the prime factorizations of these two numbers in a Venn Diagram.

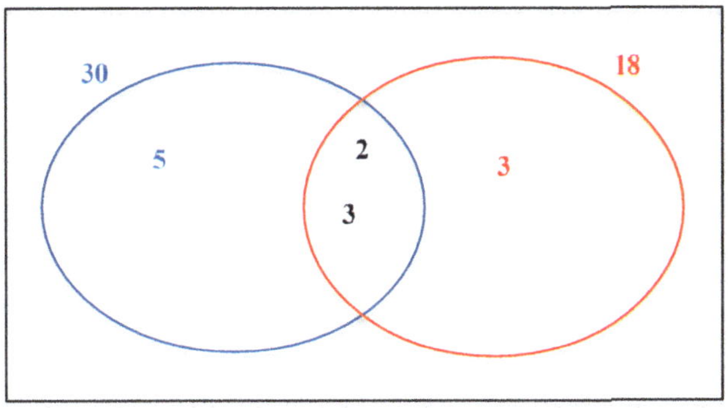

- Everything in the **blue circle** is a factor of **30.**
- Everything in the **red circle** is a factor of **18.**
- Everything in the **intersection** of the two circles (the black numbers) are **common factors.**

Step 2: Calculating the GCF

Multiply the black numbers together.

GCF of 30 and 18 is 2 x 3 = 6

Step 2: Calculating the LCM

Multiply all the black, blue & red numbers together.

LCM of 30 and 18 is 2 x 3 x 3 x 5 = 90

Use the techniques from the previous page to find the GCF and LCM of each pair of numbers.

1. 60 and 80

2. 56 and 120

3. 40 and 80

4. 36 and 144

5. 27 and 117

6. 2 and 128

7. 15 and 25

8. 18 and 76

9. 80 and 100

Middle School Math Quest

Divisibility Rules

Divisibility rules are powerful tools that can help you with prime factorization, fraction operations and many real-world applications. We'll give you a list of divisibility rules and then ask you to practice using them. If you have time, you should memorize these rules.

Divisibility Rules

A Number is Divisible by...	If...
2	**The number is even.** -352 is even so it's divisible by 2. -421 is odd so it's not divisible by 2.
3	**The sum of the number's digits is divisible by 3.** -243 is divisible by 3 because 2+4+3=9 (which is divisible by 3.) -244 is NOT divisible by 3 because 2+4+4=10.
4	**The last 2 digits are divisible by 4.** -9,824 is divisible by 4 because 24 is divisible by 4.
5	**The number ends in 0 or 5.** -1,115 is divisible by 5 because it ends with a 5.
6	**The number is divisible by 2 and 3.** (Note: the prime factorization of 6 is 2 x 3.) -432 is divisible by 6 because it's even and 4+3+2=9.
9	**The sum of the number's digits is divisible by 9.** -981,243 is divisible by 9 because 9+8+1+2+4+3=27.
10	**The number ends in 0.** -1,234,567,890 is divisible by 10 (ends with a 0).
Other Numbers	**A larger number has a smaller number as a factor if the larger number includes the smaller number's prime factorization in its prime factorization.** - 182 (2 x 7 x 13) is divisible by 14 (2 x 7). - 165 (3 x 5 x 11) is divisible by 15 (3 x 5).

Middle School Math Quest

Fill in each box with a Y or N to show whether the numbers in the left-hand column are divisible by the divisors in the top row. The first row is already completed.

A good strategy to help you memorize the rules on the previous page is to pick a column, review the divisibility rule associated with it and then try to complete the column without looking at the rule. You'll have these rules memorized in no time!

Divisibility Practice

Number	Divisors							
	2	3	4	6	9	15	18	24
60	Y	Y	Y	Y	N	Y	N	N
72								
144								
300								
360								
163,536								
18,024								
32,303								
4,664								
165								
3,280								
15,824								
162,150								
2,940								
28,652								
13,411								
2,793								
148,264								
19,782								
6,435								

Interesting Fractions

We're going to introduce you to interesting fractions now that you're a master.

Working with Fractions in Fractions

Example: Simplify $\dfrac{\frac{1}{2}}{\frac{1}{4}}$

Solution: We know the fraction bar is another way of expressing division so we can re-write our complicated fraction as $\frac{1}{2} \div \frac{1}{4} = \frac{1}{2} \times \frac{4}{1} = \mathbf{2}.$

Now, you solve some fractions in fractions:

1. $\dfrac{\frac{1}{3}}{\frac{1}{2}}$
2. $\dfrac{\frac{1}{6}}{\frac{2}{3}}$
3. $\dfrac{\frac{1}{4}}{\frac{3}{4}}$
4. $\dfrac{\frac{6}{8}}{\frac{1}{2}}$
5. $\dfrac{\frac{3}{2}}{\frac{1}{4}}$

Adding and Subtracting Fractions in a Numerator or Denominator

Example: Simplify $\dfrac{\frac{1}{2}+\frac{3}{4}}{2}$

Solution: Add the numerator terms first and then simplify as you did above:

$$\frac{\frac{1}{2}+\frac{3}{4}}{2} = \frac{\frac{2}{4}+\frac{3}{4}}{2} = \frac{\frac{5}{4}}{2} = \frac{5}{4} \div 2 = \frac{5}{4} \times \frac{1}{2} = \frac{5}{8}$$

Solve these fractions with addition and subtraction in the numerators or denominators.

6. $\dfrac{\frac{1}{3}+\frac{3}{4}}{\frac{1}{2}}$

7. $\dfrac{\frac{1}{6}}{\frac{2}{3}+\frac{5}{6}}$

8. $\dfrac{\frac{7}{8}-\frac{1}{4}}{\frac{3}{4}}$

9. $\dfrac{\frac{6}{8}}{2-\frac{1}{2}}$

10. $\dfrac{\frac{3}{2}-\frac{5}{6}}{\frac{1}{4}}$

After that practice, you're ready to solve these challenge problems.

11. $\dfrac{\frac{7}{8}-\frac{1}{4}}{\frac{3}{4}-\frac{5}{4}}$

12. $\dfrac{1-\frac{1}{4}}{1-\frac{1}{2}}$

13. $\dfrac{12-3\frac{1}{4}}{16-\frac{5}{2}}$

14. $\dfrac{1-\frac{1}{4}+3+\frac{1}{2}}{\frac{5}{6}-\frac{1}{2}}$

15. $\dfrac{\frac{1}{2}}{4(\frac{1}{8}-\frac{5}{12})}$

16. $\dfrac{6^2-3\left(\frac{7}{9}+\frac{5}{6}\right)+2\times 5}{\frac{1}{2}-\frac{3}{4}}$

Super Challenge Problem

$$\frac{1}{1+\frac{1}{1+\frac{1}{1+\frac{1}{2}}}}$$

Middle School Math Quest

Integer Intrigue

What are integers? Let's dive in and discover what they're all about.

What are number classifications?

When you first started talking, you also started counting. When we count objects, "1, 2, 3, 4, …." we are using the set of numbers that is called **natural numbers** (also known as **counting numbers**).

When you learned about the concept of zero ("If you start with 3 cookies and your brother takes away all 3 cookies, how many cookies do you have left?") and added that to your natural numbers, you started using the set of numbers called **whole numbers**.

Now, along with the whole numbers you've been using, you're going to learn about negative numbers. The set of whole numbers and negative counting numbers makes a set of numbers called **integers**.

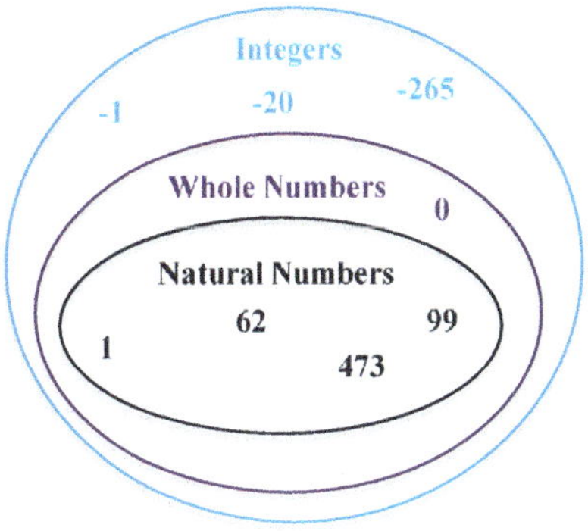

When are negative numbers used?

✓Negative numbers are used to represent direction.
"I hiked down the hill 50 feet" → **Integer is** -50

✓Negative numbers are used to represent owing money.
"I owe my dad $10 for candy" → **Integer is** -10.

✓Negative numbers are used to represent temperature below zero
"20 degrees below 0" → **Integer is** -20

Integers on a Number Line

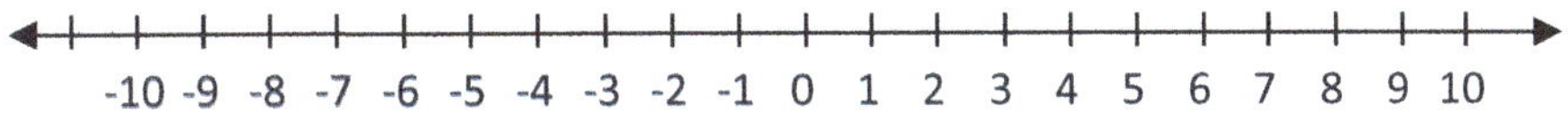

✓Negative integers are to the left of zero.

✓Negative integers are the opposite of their positive partners.

"-5 is the opposite of 5" "4 is the opposite of -4"

Absolute Value of a Number

Definition: The **absolute value** of a number is its distance from 0.

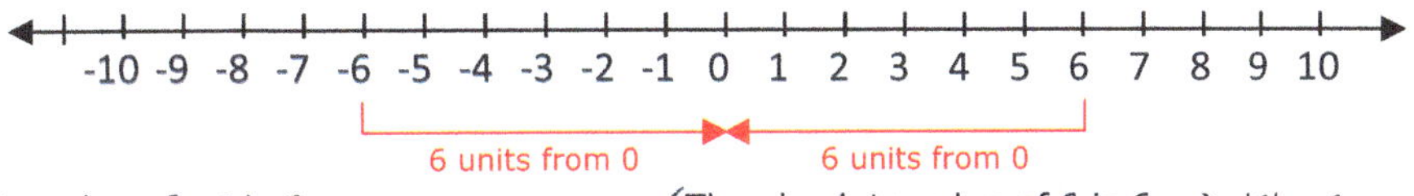

✓The absolute value of -6 is 6.
Using math symbols, we write it as $|-6| = 6$.

✓The absolute value of 6 is 6. → $|6| = 6$
-6 and 6 have the same absolute value. They are the same distance from 0.

Middle School Math Quest

Answer these questions to test your knowledge.

1. What's the opposite of -2?

2. What's the opposite of 20?

3. $|-10|$

4. $|7|$

5. $|-16|$

6. $-|-3|$

Absolute value is the same as parentheses. Operations inside parentheses are done first.

Place each group of numbers in order from least to greatest.

7. $-3,\ |-2|,\ 1,\ |-3|,\ 0$

8. $-10,\ |-12|,\ -7,\ -8,\ -|-4|$

Answer these challenge questions.

9. If the temperature is 2 below zero and then rises 10 degrees, what temperature will the thermometer show?

10. If you are on top of a 5,000-foot mountain and you climb down 2,000 feet, what integer could you use to represent the change in your altitude?

11. A shark is swimming 25 feet below sea level (think of sea level as being represented by the integer 0) and a whale is 20 feet below the shark. What integer can be used to represent the position of the whale with respect to sea level?

Adding Integers

We're going to use green team members to represent positive integers and red team aliens to represent negative integers. We'll tell you a story to help you understand.

Example 1: Add 5 + 4

5 Green Team + 4 Green Team = 9 Green Team

Story: We can think of 5 and 4 as being on the same team (positive integers). When they're brought together, they add to each other.

$$5 + 4 = 9$$

Example 2: Add $-3 + -2$

3 Red Aliens + 2 Red Aliens = 5 Red Aliens

Story: -3 and -2 are on the same team (negative integers). When they're brought together, they add to each other.

$$-3 + -2 = -5$$

Example 3: Add $-3 + 5$

3 Red Aliens + 5 Green Team = 3 Red Aliens & 5 Green Team

Story: What happens now? We've got reds and greens grouped together. But aliens and green team people don't like each other. So, they cancel each other out.

Battle of the Teams

3 (Negative) Aliens & 5 (Positive) Greens = 2 (Positive) Greens

3 (negative) aliens cancel out 3 (positive) greens. We're left with 2 (positive) greens.

Story: When you bring more (positive) greens than (negative) aliens to the playing field, the (positive) greens win because they have more players.

$$-3 + 5 = 2$$

Example 4: Add $-4 + 1$

4 Negative Aliens + 1 Green Team = 3 Negative Aliens

Story: When more (negative) aliens than (positive) green members meet up, the (negative) aliens win because they have more players.

$$-4 + 1 = -3$$

Adding and Subtracting Integers

We explored examples of adding integers on the previous pages. Now, we'll give you algorithms that will help you add integers together without using aliens.

Algorithms for Adding Integers

When the Signs on the Numbers Match

Step 1: Add the absolute value of the numbers together.

Step 2: Match the sign of your answer to the sign of the numbers.

Ex 1: Find $12 + 15$

$12 + 15 = 27$

$12 + 15 = 27$

Ex 2: Find $-7 + -13$

$7 + 13 = 20$

$-7 + -13 = -20$

When the Signs on the Numbers Don't Match

In integer addition, if the 2nd number is negative, we put parentheses around it to separate it from the addition sign.

Step 1: Find the absolute value of each number.

Step 2: Find the difference of the absolute values (largest – smallest).

Step 3: Match the sign of your answer to the sign of the number with the greater absolute value (also known as its **magnitude**).

Ex 3: Find $612 + (-524)$

$|612| = 612$ Largest
$|-524| = 524$

$612 - 524 = 88$

$612 + (-524) = 88$

Answer is positive because 612 has the greatest absolute value.

Ex 4: Find $-323 + 147$

$|-323| = 323$ Largest
$|147| = 147$

$323 - 147 = 176$

$-323 + 147 = -176$

Answer is negative because -323 has the greatest absolute value.

Algorithm for Subtracting Integers

Rule: Subtraction is the same as adding the opposite of a number.

Since subtraction can be converted to addition, we convert subtraction problems to addition problems and use our addition algorithms to add integers.

Ex 5: Find $5 - 3$
We're subtracting 3 from 5. That's the same as adding the opposite of 3 to 5.

$$5 - 3 = 5 + (-3) = 2$$

Ex 6: Find $-10 - (-3)$
We're subtracting -3 from -10
That's the same as adding the opposite of -3 to -10.

$$-10 - (-3) = -10 + 3 = -7$$

Middle School Math Quest

Practice your integer addition and subtraction skills with these problems. Don't forget to convert subtraction problems to addition problems.

12. $-406 + (-284)$

13. $-130 - (-275)$

14. $-135 - 408$

15. $170 + (-340)$

16. $-451 + 443$

17. $169 + (-410)$

18. $-325 + (-177)$

19. $275 + (-480)$

20. $-192 - 199$

21. $-3 + -2 + -5$

22. $-7 + 7 + -10$

23. $-26 - 12 + 17$

24. $8 - 16 - 23$

25. $-97 - 6 + 11$

26. $66 + (-72) - (-14)$

27. $100 - (-9) - 18$

28. $-25 + -25 + 50$

29. $-1 - (-1) + (-1)$

30. $36 + -24 - (-49)$

Multiplying/Dividing Integers

Rule: When **multiplying integers**, first multiply the digits then determine the final sign. Remember, negative signs stand for 'the opposite' of a number. When multiplying, remember that two negatives make a positive.

Ex 1: $-2 \times 3 = -(2 \times 3) = \mathbf{-6}$ **Ex 2:** $-2 \times -3 = -(-(2 \times 3)) = -(-6) = \mathbf{6}$

Ex 3: $-1 \times -1 \times -1 = -(-(-(1 \times 1 \times 1))) = -(-(-1)) = \mathbf{-1}$

Integer Division follows the same rules. First, divide your digits and then work through your negative signs.

Ex 4: $-16 \div -4 = -(-(16 \div 4)) = -(-4) = \mathbf{4}$ **Ex 5:** $\frac{-20}{2} = -(\frac{20}{2}) = \mathbf{-10}$

Pattern Rule: When multiplying or dividing integers, count the number of negative signs in the problem. If there is an **even number of negatives** (0, 2, 4, ...), your answer is **positive**. If there is an **odd number of negatives** (1, 3, 5, ...), your answer is **negative**. This only works when there is no addition or subtraction.

Integers and Exponents

Question: Is there a difference between -2^2 and $(-2)^2$?

Answer: Yes, there is a difference. You need to think carefully about order of operations when you work with negative numbers and exponents.

Ex 6: $(-2)^2 = (-2) \times (-2) = -(-(2 \times 2)) = -(-4) = \mathbf{4}$

Exponents first. Here, the exponent is attached to the parentheses, so you are multiplying everything inside the parentheses twice (because the exponent is a 2).

Ex 7: $-2^2 = -(2)^2 = \mathbf{-4}$

Exponents first. The exponent is right next to the base of 2, so we only square 2. Then we apply the negative sign afterwards. You can think of the negative sign as multiplying by -1. Multiplication is done after exponents.

Middle School Math Quest

Try these problems. Start each problem thinking about your integer rules and any order of operations you should apply.

31. $-3 \times (-4) - 13$

32. $-4^2 - 4 + (-7)$

33. $\frac{32 \times (-3)}{-6}$

34. $-5 - \frac{-3 \times (-6)}{-2}$

35. $-3^3 - 2 \times (-3)$

36. $10 - (-2) \times (-4)$

37. $(-5)^2 - (-8)$

38. $(-5 - 7)^2 + -6$

39. $6 - (-2) \times (-8) \div (-4)$

40. $-13^2 - (-11)^3$

That was a tough topic!

Greens versus Aliens

Now that you've practiced working with integers, we're going to introduce you to a cool game called Greens versus Aliens. We took our inspiration from John Conway's fantastic Game of Life (search it up online). Like Conway's game, our game is a cellular automaton; something, a machine or process, that automatically continues based on rules that are given at the beginning.

We'll walk you through a few examples so you understand this world. Then, you'll get to explore fun worlds we start for you. **Greens** will represent **positive integers** and **Aliens** will represent **negative integers**.

Greens = 1 and **Aliens = -1**

Step 1: Start by placing players on a given grid. You get to decide how many players you place on the grid.

Initial Grid → **Players Added to the Grid**

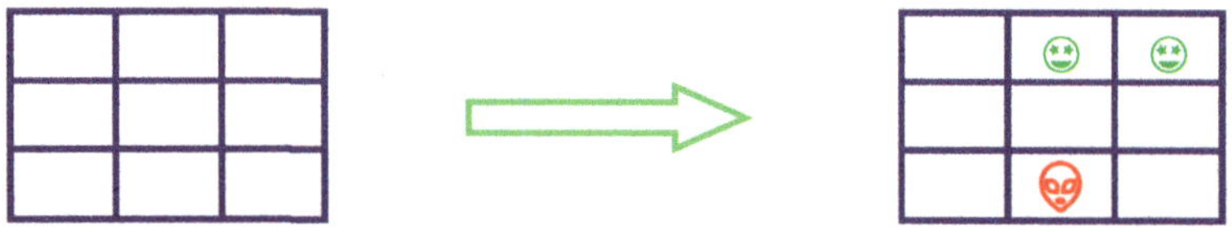

Step 2: For each cell, sum the values of the surrounding neighbors' cells.

EXAMPLE 1

Players Grid

Sums Table

1	1	1
0	1	1
-1	0	-1

The yellow cell has three neighboring (orange) cells. Only one of those cells has a **Green** in it so we enter 1 in our Sums Table for the yellow cell. Remember, a **Green** player represents **1**. We don't include the yellow cell's **Green** player in in our sum, only the neighbors' **Green**.

EXAMPLE 2

Players Grid

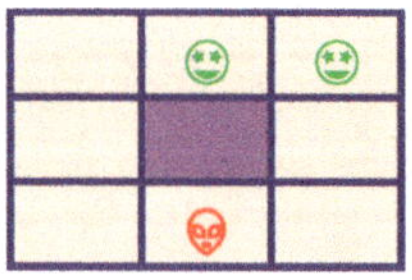

Sums Table

1	1	1
0	1	1
-1	0	-1

The pink cell has eight neighboring (orange) cells. Two cells have a **Green** player and one cell has an **Alien** player so we add $1 + 1 + (-1) = 1$, our sum for that cell.

Step 3: For each cell, find the product of any neighbors. If all the neighbors' cells are empty, enter 1 in that cell.

EXAMPLE 1

Players Grid

Products Table

1	1	1
-1	-1	-1
-1	1	-1

The yellow cell has five neighboring (orange) cells. One cell has a **Green** player and another cell has an **Alien** player so we multiply $1 \times (-1) = -1$, our product for that cell.

EXAMPLE 2

Players Grid

Products Table

1	1	1
-1	-1	-1
-1	1	-1

The blue cell has five neighboring (orange) cells. All are blank so we enter a 1 in the products table for that cell.

Step 4: Decide whether 'living' cells (those with players in them) will remain alive.

A living cell remains **alive** if the absolute value of the cell sum is 2 or 3 (from Step 2). If |cell sum| doesn't equal 2 or 3, the cell will die. If the cell remains living, it will remain on the same team if the product of its neighbors is 1. If their product is -1, the cell will switch teams in the next generation.

Generation 0

Player Grid

Sums Table	1	1	1
	0	1	1
	-1	0	-1
Products Table	1	1	1
	-1	-1	-1
	-1	1	-1

Apply living cell rule.

Generation 1

Player Grid			

All cells that were alive in our initial generation will die off in Generation 1. The two **Green** cells each have only one living neighbor and that's not enough community support. The **Alien** cell will die off because it had 0 living neighbors. Since each of these 3 cells didn't have enough living neighbors (tallied in the Sums table), we didn't even have to look at the Products table.

Sometimes, we start a colony (with Generation 0) and it will die off completely by Generation 1. In our case, though, we have one more step to follow…

Step 5: Decide whether empty cells will turn on in the next generation.

An empty cell changes to **alive** in the next generation if it has a sum of **-1** or **1** living neighbors. It will decide which team to be on (**Greens** or **Aliens**) based on the product of its neighbors.

Generation 0

Player Grid		🙂	🙂
		👽	
Sums Table	1	1	1
	0	1	1
	-1	0	-1
Products Table	1	1	1
	-1	-1	-1
	-1	1	-1

Apply empty cells rule. →

Generation 1

Player Grid	🙂		
		👽	👽
	👽		👽

The yellow cell turns on in Generation 1. It was empty at the start of this colony and it was persuaded to turn on by the **Green** player to its right. The blue cell was also empty in Generation 0. It turned on because the sum of its living neighbors was 1 $[1 + 1 + (-1) = 1]$. It's an **Alien** because the product of its Gen 0 3 living neighbors is -1.

Now that we have Generation 1, what do we do? We go through the Steps 2 - 5 again to determine what Generation 2 will be. Here's a summary of our steps.

Greens v Aliens Rules of Play

Step 1: Start by placing players on a grid. You get to decide how many players you place on the grid. This step is only done for Generation 0.

Step 2: For each cell, sum the values of the surrounding neighbors' cells.

🙂 **Greens = 1** and 👽 **Aliens = -1**

Step 3: For each cell, find the product of any neighbors. If all the neighbors' cells are empty, enter 1 in that cell.

Step 4: Decide whether 'living' cells (those with players in them) will remain alive.

A living cell remains **alive** if the absolute value of the cell sum is 2 or 3 (Step 2). If |cell sum| doesn't equal 2 or 3, the cell will die. If the cell remains living, it will remain on the same team if the product of its neighbors is 1. If their product is -1, the cell will switch teams in the next generation.

Step 5: Decide whether empty cells will turn on in the next generation.

An empty cell changes to **alive** in the next generation if it has a sum of **-1** or **1** living neighbors. It will decide which team to be on (**Greens** or **Aliens**) based on the product of its neighbors.

Now that we've summarized the rules, we'll show you what Generations 2 and 3 look like based on our current Generation 1.

Generation 1

Player Grid	🙂		
		👽	👽
	👽		👽
Sums Table	-1	-1	-2
	-1	-2	-2
	-1	-4	-2
Products Table	-1	1	1
	1	-1	1
	-1	1	1

Generation 2

Player Grid		🙂	
	🙂	🙂	👽
			👽
Sums Table	3	1	1
	2	0	1
	2	0	0
Products Table	1	-1	-1
	1	1	-1
	1	1	-1

Generation 3

Player Grid			👽
	🙂		
Sums Table	1	0	0
	0	0	-1
	1	1	0
Products Table	1	-1	1
	1	-1	-1
	1	1	1

Challenge 1: It's your turn now. Fill in Generations 4 through 6 (below) based on the information above. If you don't want to draw **Greens** or **Aliens** you could use **Os** for **Greens** and **Xs** for **Aliens** (mathematicians frequently simplify problems by using easier representations.)

Generation 4

Player Grid			
Sums Table			
Products Table			

Generation 5

Player Grid			
Sums Table			
Products Table			

Generation 6

Player Grid			
Sums Table			
Products Table			

Challenge 2: Determine Generation 3 for each of these colonies. Create your own Sums and Products tables on a separate piece of paper.

Colony A Gen 0

	🙂	👽
🙂		

Colony B Gen 0

	👽	

Looks like you're ready for the big challenges now. We're going to give you some expanded grids to practice your **Greens** vs **Aliens** game skills.

Challenge 3: Determine Generation 3 for each of these 4x4 colonies. Don't forget to calculate Sums and Products Tables for each generation.

Colony C
Gen 0

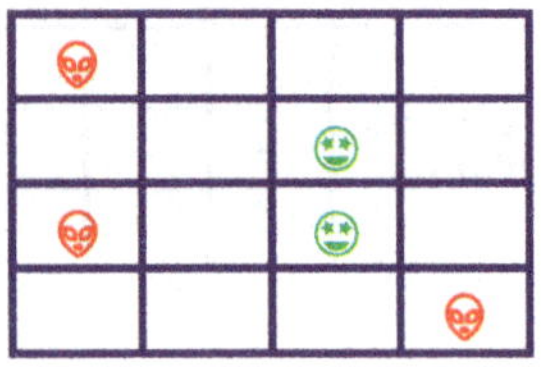

Colony D
Gen 0

Colony E
Gen 0

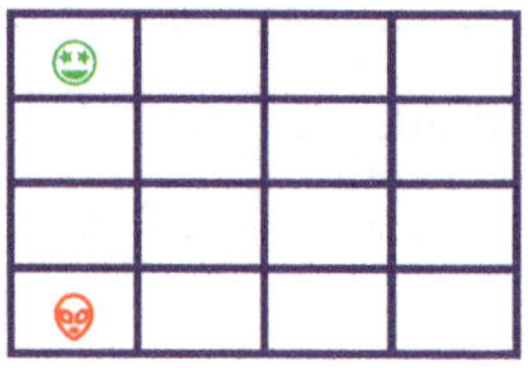

Challenge 4: For your final challenge, you'll work with 5x5 grids and find Generation 3 for each colony.

Note that Colony F starts with the same players as Colony C – it just provides one more row and one more column. Will Generation 3 of Colony F be the same as Colony C's generation 3?

Colony F
Gen 0

Colony G
Gen 0

You got some great practice playing **Greens** versus **Aliens**. We'll re-visit this world (with different rules) in our Super Stats section. There, you'll get to work with larger numbers.

Positively Negativity

This might be the hardest page in the entire quest book! Use your order of operations, exponents and integer skills to complete these challenges. These are designed to stump the hardest-working student so don't be discouraged if you don't get the correct answer on your first try.

1. $-3-(-2)^2-15$

2. $\dfrac{-7\times(-2)-(14\times2)}{-3}$

3. $-5-3(-18-2)$

↑ A number next to parentheses means multiply whatever is inside the parentheses by the number on the outside

4. $-3^2-(-3)^3-(-4-7)^2$

5. $-1-(1-1-1)\times(-1-1^2)$

6. $-\frac{1}{2}-\left(-\frac{1}{4}\right)-\frac{1}{8}-\left(-\frac{1}{16}\right)$

7. $-1\times(-2)-3(-4\times5-6)$

8. $-3\frac{1}{2}-\frac{1}{2}[-2-(-3)]$

9. $-2-2^2\times(-2^3)$

10. $-10-9[-8-(-7)]$

11. $-1-1.2-(-1.4-1.6)$

12. $-6\times(-4)\times\left(-\frac{1}{3}\right)\times(-2)$

Middle School Math Quest

Super Stats

You may have already learned how to calculate the mean, median, and mode of a data set. But, some schools don't cover these topics until 6th grade, or later, so we'll review the topics and give you some cool challenges

Mean, Median and Mode Summary

Example Dataset: 6, 5, 3, 7, 4, 1, 2, 8, 11, 9, 10, 6, 12

Mean (Average)

Add all the observations together and divide by the number of observations.

Add 6 + 5 + 3 + 7 + 4 + 1 + 2 + 8 + 11 + 9 + 10 + 6 + 12.

These observations sum to 84.

We divide this total by the number of observations:
$(84 \div 12 = 7)$

The mean, or average, is 7 for this dataset.

Median

Sort the data from lowest to highest.

1, 2, 3, 4, 5, 6, 6, 7, 8, 9, 10, 11, 12

Middle Observation

Determine the middle number. That's the median.

The median is 6.

Special Case: If there is an even number of observations, take the average of the two middle observations.

Ex: 1, 2, 3, 4, 5, 6

Middle of the Dataset

The middle is between 3 and 4 so our median is 3.5.

Mode

The observation, or observations, that appear the most in the data set is the mode.

6 appears twice in the dataset while every other value only appears once.

The mode is 6.

Special Case: If all values only appear once in the dataset, we say the dataset does NOT have a mode.

Special Case: You can have more than one mode. If we had a second 12 in our dataset, we would have two modes, 6 and 12.

You are a super statistician!

We'll work with these statistics by using a cellular automaton similar to **Greens** v. **Aliens**. Make sure you work through those challenges before you try these.

Stats State Rules of Play

Step 1: You'll be given **green** and **red** numbers on an initial grid (or you can create your own colony by placing numbers in a grid.) Think of these numbers as multiple team members from the **Greens** v **Aliens** game. This is only done for Generation 0.

Step 2: For each cell, calculate the **average** of the neighboring cells. Don't include cells that are blank. They aren't included in the sum and they aren't counted as an observation.

Step 3: For the entire current generation, calculate the **median** of the population. Again, don't include cells that are blank in your calculation.

Note: For both Steps 2 and 3, if a cell has 0 in it, you **WILL** include it in your calculations for **average** and **median**.

Step 4: Decide whether 'living' cells (those with players in them) will remain alive.

A living cell remains **alive** if the **average** of the cell's neighbors, from Step 2, is no more than one away from the cell's value. If a living cell doesn't match this criteria, it will die in the next generation.

Cells are encouraged to remain alive when they're close to the **average** of their neighbors.

Step 5: Decide whether empty cells will turn on in the next generation.

An empty cell changes to **alive** in the next generation if the **average** of the cell's neighbors, from Step 2, is 2 or more away from the population **median**. If a cell changes to **alive** in the next generation, its value will be the **average** of the neighbors, rounded to the nearest whole number.

Cells surrounding the empty cell encourage it to turn on when their **average** is significantly different from the **median**.

Next up? An example to help you understand the rules. Then, you'll get to do your own work with interesting Stats States.

Stats State Example

Generation 0

Colony	-5		
	-4		
	0	1	
Average	-4	$-4\frac{1}{2}$	
	$-\frac{4}{3}$	-2	1
	$-\frac{3}{2}$	-2	1

Median: -2

Generation 1

Colony	-5	-5	
			1
			1
Average	-5	-2	-2
	-5	-2	-2
		1	1

Median: -2

Generation 2

Colony	-5		
	-5		
		1	1
Average	-5	-5	
	-2	-2	1
	-2	-2	1

Median: -2

The green cell doesn't have an average because it doesn't have any living neighbors.

The orange cell shows an average that can't be represented as a decimal. It's better to use mixed numbers and fractions for averages in this activity.

The blue cell turns on in Generation 1. Its corresponding average in Generation 0 was $-4\frac{1}{2}$. With negative numbers, think of rounding the number first and then applying the negative sign. Since we would round $4\frac{1}{2}$ to 5, we'll round $-4\frac{1}{2}$ to -5.

In all 3 generations, our median is -2. It's unusual to have the same median across multiple generations.

In Generation 1 and Generation 2, we have two modes, -5 and 1. These observations both show twice in each colony.

Now it's your turn. Calculate Generation 2 for these initial Stats States. After you finish each Stats State, list the modes (if they exist) for each generation. Don't forget to calculate an Average Table for each generation, and a median to help you follow the Stats State rules.

State 1, Gen 0

1		
	5	

State 2, Gen 0

	7	-4
1		-15

State 3, Gen 0

30	13	-30
20	15	

Here are some tougher challenges. Determine Generation 2 for each of these 4 x 4 initial states.

State 4, Gen 0

	1		
		5	

State 5, Gen 0

1		1	
	7		7
1		1	
	7		

State 6, Gen 0

	-4		
	7		

State 7, Gen 0

2		4	
	6		8
-2		-4	
	-6		-8

We love Stats State. If you write a program to simulate these rules, you'll make amazing discoveries. For example, in State 5, we can add one more number to the initial generation and the entire colony will disappear by Generation 1.

Let's use your mean formula and apply it to more complicated problems. Don't forget – Add up all observations and then divide by the number of observations to calculate the mean/average.

Mean/Average Formula

$$\frac{x_1 + x_2 + \cdots + x_n \text{ (Sum of all Obs)}}{n \text{ (\# of Obs)}} = \textbf{Average}$$

You can use this formula and variables for your unknowns to solve the following problems. Others may require a little bit of mathematical thinking. Don't be discouraged if you find the problems on the next page hard. Think about each one and write down what you know. You can always use the answer key to help figure these out.

1. You have a dataset with the values 22, 27, 15, 14, and x. What is x if the average is 20?

Using our formula, we have: $\frac{22+27+15+14+x}{5} = 20 \rightarrow \frac{78+x}{5} = 20 \rightarrow 78 + x = 100 \rightarrow x = 22$

2. If Max had an average of 55 on his first 5 math tests, what is the minimum score he'll need to get on his 6th test in order to have a passing average of 60?

3. What is the average of the first six odd numbers?

4. Susan scored 1, 2, 3, 4 and 5 goals in her first 5 soccer games. Then in her 6th and 7th games, she scored enough that her average goals per game is now 5. How many total goals did she score in her 6th and 7th soccer games?

5. What is the average of 1, 2, 3, 4, 5, 6, 7, 8, 9, 10 and 11? How is it related to the median of the dataset?

6. What is the average of 1, 2, 3, ..., 47, 48, 49, 50, and 51? You should be able to use learnings from Q5 to answer this!

7. What is the average of the even numbers from 2 through 100? How is this different from what you saw in Q9 and Q10?

8. You have 5 numbers representing the daily high temperature in the winter in Yakutsk, Russia. Their average is -2. Then, on the 6th day, the temperature plummets and your average is now -10. What was the high temperature on Day 6 in Yakutsk?

9. In a set of 10 integers, the average of the smallest two integers is 10. The average of the three smallest integers is 11, the average of the four smallest integers is 12.... The pattern continues with all 10 integers. What is the median of the dataset?

10. You have 3 numbers and two of them are $\frac{3}{4}$ and $\frac{5}{6}$. What is the third number if the average of all three numbers is 3?

BONUS SHEETS

The following pages include challenging, fun puzzles you will enjoy.

You can do some of the work when you first get this packet but you'll need to practice all your skills before you can solve everything.

Good Luck!

Middle School Math Quest

Cross Number

DOWN

1 6^2
2 7,095 ÷ 5
3 2,306 x 4
4 One more than 8-across.
5 The same as 9-down.
7 7 x 12
9 232 ÷ 8
11 The first 2 digits are 968 ÷ 11. The last digit is one more than the first digit.
12 The first digit is the sum of the middle digit and the last digit. The first digit is 3^2.
14 The first digit and last digit are the same. The first digit is twice as big as the middle digit.
15 One hundred more than 16-across.
19 12,414 – 6,451
20 The last two digits are three more than the first two digits. The first digit and the third digit are the same.
21 $\frac{1}{10}$ of 590
23 14 x 10 ÷ (2 + 2)
24 One more than half of 5-across.
26 One less than three times greater than 8-across.
28 378 ÷ 18

ACROSS

1 1,053 ÷ 3
3 The first two digits are 14 x 7. The last digit is $\frac{3}{4}$ of 4.
5 Ten less than 1-down.
6 The last digit is half of the first digit. The middle digit is two times greater than the first digit.
8 2 x 24-down + 4
10 8 x 8 + 39 x 2
11 The last 2 digits are the same as 25-across. The first digit is one less than the middle digit.
13 The sum of the first digit and the middle digit is equal to the last digit. The first digit and the middle digit are the same.
16 2,453 less than 20-down.
17 The middle digit is four times greater than the first digit. The last digit is the third prime number.
18 25 x 39
20 The last digit is half of the middle digit.
22 (168 + 65) x 4
25 9 x 11
27 The middle digit is the third prime number. The last digit is the second prime number.
28 768 ÷ 32
29 The middle digit is $2\frac{2}{8} - 2\frac{9}{36}$. The last digit is two less than the first digit.
30 The first digit – the last digit = the middle digit.

Secret Codes

Use your mathematical thinking and reasoning to solve these secret codes.

Double Trouble

(Hint: What do you know about digits that are twice as large as other digits?)

1. It is more than 2 times 60,000 and less than 2 times 450,000.

2. The tens digit is 2 times the ones digit.

3. The hundreds digit is 2 times the tens digit.

4. The sum of the hundred thousands, ten thousands, and thousands digits is twice as large as the sum of the hundreds, tens, and ones digits.

5. The ten thousands digit is two more than the thousands digit.

6. The hundred thousands digit is twice as large as the thousands digit.

A Little Odd

(Hint: Do you have an even or an odd number when you add two odd numbers together?)

1. It is a number between 13,579 and 97,531.

2. The number is odd.

3. An odd number of digits are odd.

4. The ten thousands digit is an odd multiple of the ones digit.

5. The thousands digit is the sum of the ones digit and the first positive odd counting number.

6. The ones digit is three times the tens digit.

7. The ten thousands digit is either one more or one less than the thousands digit.

8. The hundreds digit is a multiple of 3.

Middle School Math Quest

Tower Math

Tower builders need your help! To keep these towers standing, the bricks across from each other must be the same height. The height of each brick is written on it, but they need the SAME missing number filled in on both sides to make the equations true. All heights are positive whole numbers, and all missing numbers are between 1 and 10 (inclusive). The first one has been done for you.

Hint: Think of reverse order of operations (or solving an algebraic equation).

$(1 + \boxed{4})^2$	$=$	$3 \times 7 + \boxed{4}$
$\square \div 2$	$=$	$\square - 4$
$8 \times \square$	$=$	$35 + \square$
$\square + 4$	$=$	$\square \times 3 - 2$
$\dfrac{\square - 2}{2}$	$=$	$\square - 5$
$\dfrac{\square}{3}$	$=$	$\square - \dfrac{12}{3}$
$\square^2$	$=$	$\square \times 2$
$\square^2$	$=$	$\square \times 9$

$\dfrac{\square^2}{7}$	$=$	$\square$
$\dfrac{\square + 2}{3}$	$=$	$\dfrac{\square}{2}$
$\dfrac{\square \times 7}{5}$	$=$	$\square + 4$
$\dfrac{\square}{7} + 5$	$=$	$\square - 1$
$\square \times (3 + 2)$	$=$	$\square^3 + 4$
$\dfrac{\square}{3} + 1$	$=$	$\dfrac{\square}{2}$
$\dfrac{\square \times 2}{4}$	$=$	$\dfrac{\square + 15}{5}$

Middle School Math Quest

Triangle Challenges

Fill in the blank circles so each side of the triangle has a product equal to the red target number at the top. Note: You can only use a number once in each puzzle (negative and positive numbers with the same absolute value are different numbers). There may be more than one solution.

Example

Puzzle

600

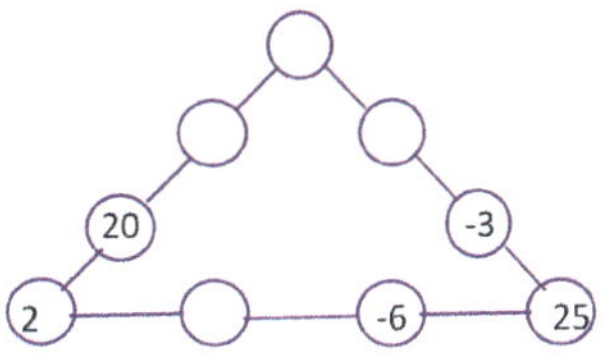

Solution

600

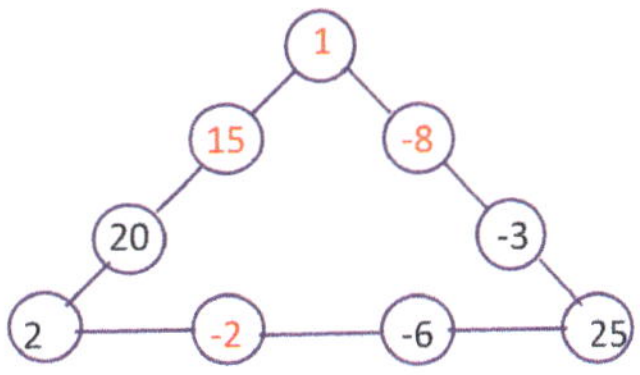

Mathematical Thinking

We start by calculating the prime factorization of 600 = 2x2x2x3x5x5. That helps us determine what is needed for each side.
The bottom needs a 2 and it must be negative so we can fill that in. Then, we look at left/right sides. The right side still needs three 2s while the left side already has its 2s. So, we'll put a -8 on the RHS (we also could have used 8). We finish from there.

-900

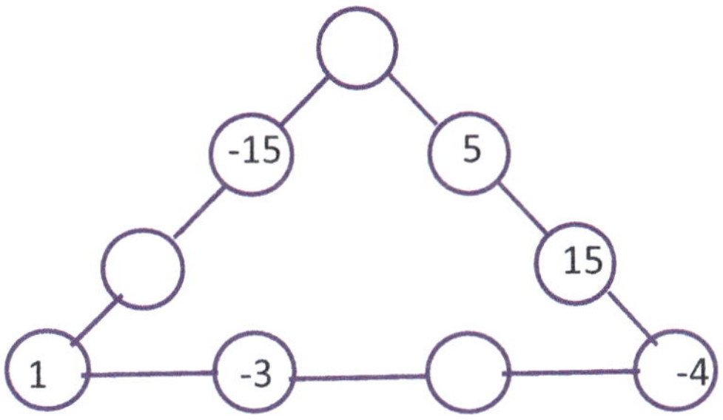

1,260

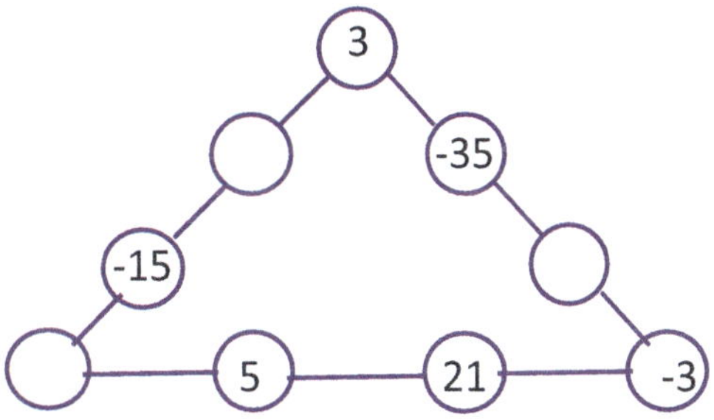

-240

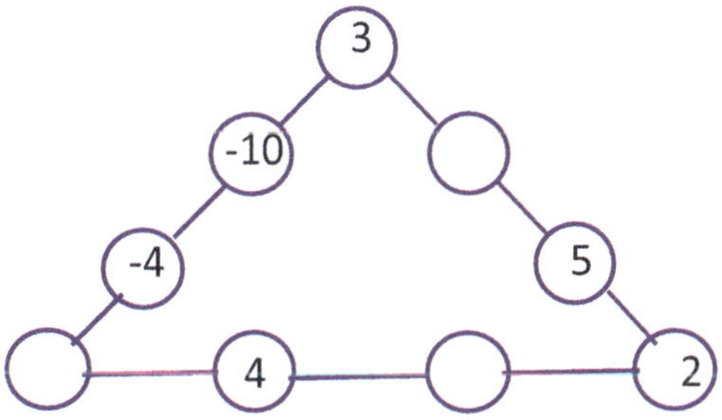

1,000

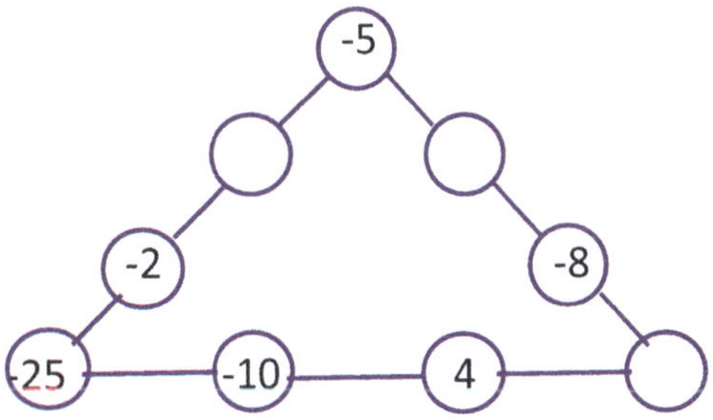

144

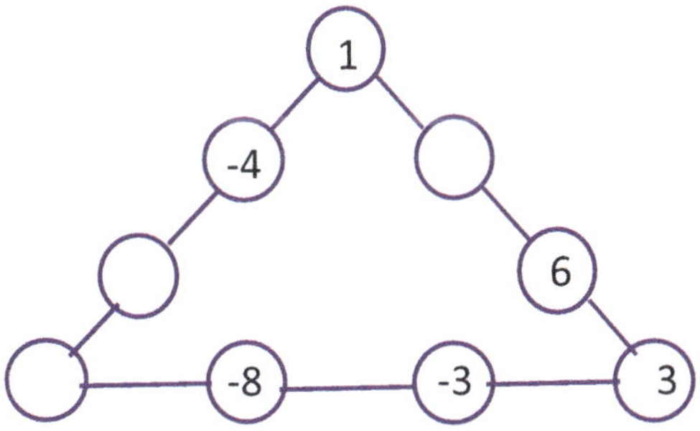

-720

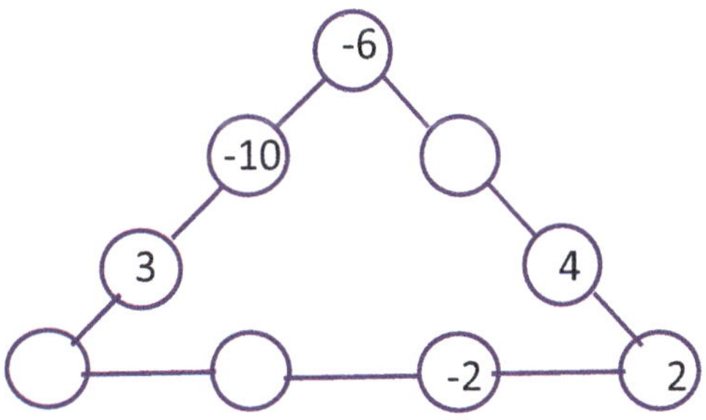

Middle School Math Quest

Multiplication and Division Quest

Do we multiply or do we divide? That's the QUESTion! Decide whether there should be multiplication or division signs between each pair of numbers. At the end, if you work from left to right, you should match the number on the right-hand side of the equals sign. There may be more than one solution.

All intermediate steps will have a whole number answer. Logic and divisibility rules can help you (Ex: In Challenge 1, 12 doesn't divide evenly into 20 so we know we must have a multiplication sign for our first step).

We've done the first one for you.

Challenge 1:	20 x	12 ÷	2 x	10 ÷	3 ÷	8	=	50

20 x 12 = 240 → 240 ÷ 2 = 120 → 120 x 10 = 1,200 → 1,200 ÷ 3 = 400 → 400 ÷ 8 = 50

Challenge 2:	34	2	12	10	4	5	=	102
Challenge 3:	14	7	12	4	3	12	=	24
Challenge 4:	40	8	8	12	8	10	=	384
Challenge 5:	30	10	6	2	5	5	=	1
Challenge 6:	26	10	10	7	13	7	=	2
Challenge 7:	36	2	2	11	3	3	=	11
Challenge 8:	27	3	7	9	7	9	=	729
Challenge 9:	15	3	8	6	8	3	=	10
Challenge 10:	17	6	6	8	6	12	=	9,792

Middle School Math Quest

Power Panel: Solving Quizzles

Quizzles can look very strange if you haven't solved them before. Here are strategies you can use to solve this Quizzle, Counting Students.

		Objects			students		
		benches	swings	Bikes	Ethan	Fred	Grace
numbers	7						
	13						
	20						
students	Ethan						
	Fred						
	Grace						

Counting Students Clues:

1. Ethan counted a prime number of swings.
2. Fred counted fewer items than Ethan.
3. Of Fred and the person who counted bikes, one of them counted 20 objects and the other counted 7.

Solving Strategies: The first thing to note is that every student, number, and object must be used once and only once. **Clue 1** tells us that Ethan counted swings, so we immediately know that Fred and Grace did not count swings and that Ethan did not count benches or bikes. We can fill this in to the grid in red, using a circle to mark pairings we know are true and an X to mark pairings we know are false.

		Objects			Students		
		benches	swings	bikes	Ethan	Fred	Grace
Number	7				X	O	X
	13				O	X	X
	20		X		X	X	O
Students	Ethan	X	O	X			
	Fred		X				
	Grace		X				

We also know from Clue 1 that Ethan counted a prime number, so Ethan could not have counted to 20. We can put an X for Ethan in the 20 row, and since we already know that Ethan is matched with swings, we also know that 20 cannot be the number of swings. We can add these two Xs to the grid.

Clue2 tells us that Fred counted fewer items than Ethan counted. This automatically tells us that Ethan didn't count the smallest number and Fred didn't count the largest number of items. Since we already know the largest number Ethan could have counted is 13, Fred must have counted 7 items and Ethan counted 13 items. This leaves Grace with 20 items.

Since Ethan counted 13 items and they were swings, we can match 13 with swings.

Finally, Clue 3 begins "Of Fred and the person who counted bikes," telling us that Fred is not the person who counted bikes. From here, practice using our strategies to fill in the rest of the grid.

		objects			students		
		benches	swings	bikes	Ethan	Fred	Grace
numbers	7		X		X	O	X
	13	X	O	X	O	X	X
	20		X		X	X	O
students	Ethan	X	O	X			
	Fred		X	X			
	Grace		X				

Answer:

Ethan	13	swings
Fred	7	benches
Grace	20	bikes

Middle School Math Quest

Divisibility Quizzle

Match the students to their favorite and least favorite numbers.

Clues:

1. Kian's favorite number is divisible by everyone else's least favorite numbers.

2. The people who have favorite numbers divisible by 5 are the person whose least favorite number is 6, the person whose least favorite number is 2, and Lena.

3. Nate's favorite number is not divisible by any of the other favorite numbers.

4. Maggie's least favorite number is not divisible by 3.

5. Of Ollie and the person whose favorite number is 25, one of them has the least favorite number 6 and the other has the least favorite number 21.

6. Maggie's favorite number is divisible by Ollie's favorite number.

		least favorite					students				
		2	3	6	21	100	Kian	Lena	Maggie	Nate	Ollie
favorite	5										
	25										
	42										
	77										
	360										
students	Kian										
	Lena										
	Maggie										
	Nate										
	Ollie										

Middle School Math Quest

Answer Key

Whole Number Operations, Calculating and Estimating (p. 9)

1. 150,297
2. 556,023
3. 597,310
4. 85 R62
5. 109 R1
6. 760 R62

7.

ROUND TO THE NEAREST...	NUMBER 59,261
TENS	59,260
HUNDREDS	59,300
THOUSANDS	59,000
TEN THOUSANDS	60,000
HUNDRED THOUSANDS	100,000

8.

ROUND TO THE NEAREST...	NUMBER 167,089
TENS	167,090
HUNDREDS	167,100
THOUSANDS	167,000
TEN THOUSANDS	170,000
HUNDRED THOUSANDS	200,000

9. $500 + 100 = 600$

10. $24{,}000 \div 400 = 60$ (Other Estimate)

11. $50 \times 10 = 500$

12. $600 - 200 = 400$ (Other Estimate)

13. $600{,}000 \div 1{,}000 = 600$

14. $200 \times 30 \div 10 = 600$

Whole Number Operations, Bar Modeling (pp. 10-11)

Challenge 1: Logan has 45 books.

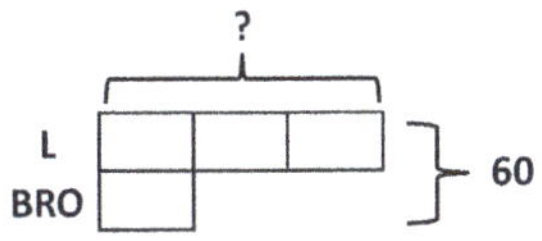

Challenge 2: Jackson has 80 pencils.

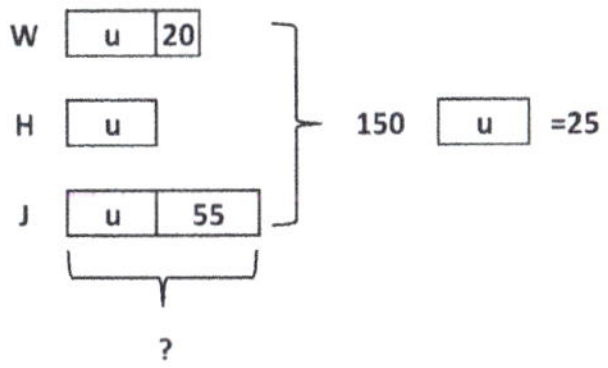

Challenge 3: Grace has 48 apples.

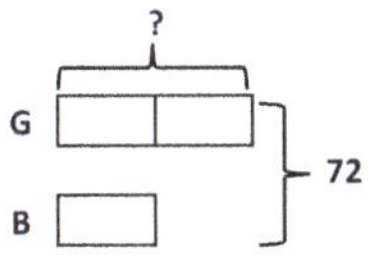

Challenge 4: A = 152, B = 79, M = 76

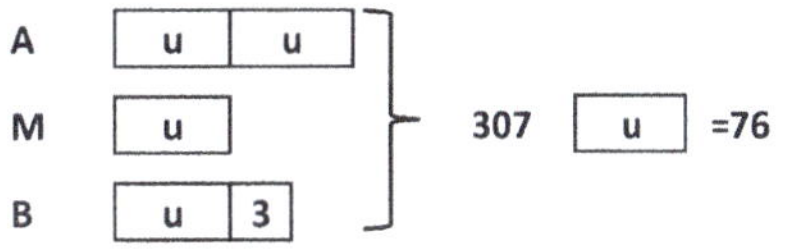

Challenge 5: 30 boys and 20 girls

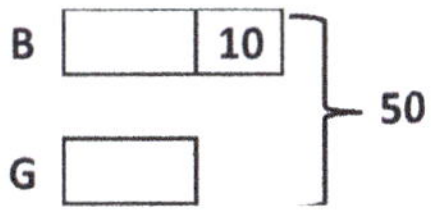

Challenge 6: There are 225 adults on the flight.

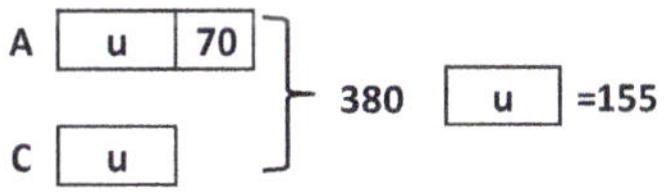

Challenge 7: Numbers are 117, 118 and 119.

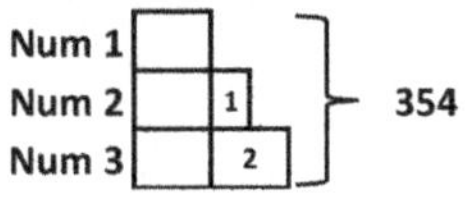

Challenge 9: \$100, \$200, \$200 and \$400

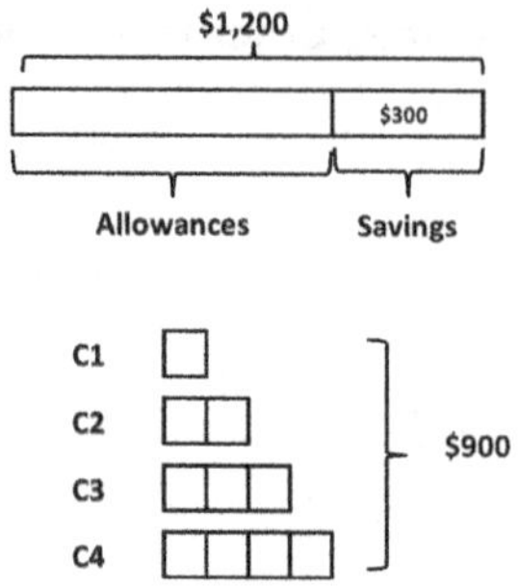

Challenge 8: Numbers are 61, 63, 65 and 67.

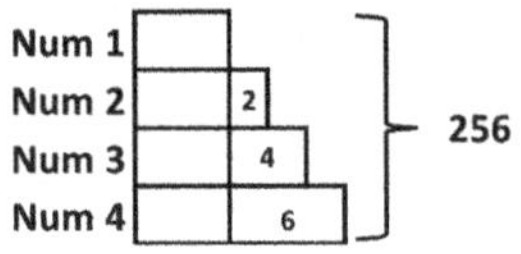

Challenge 10: Ben had \$150; Jerry had \$75

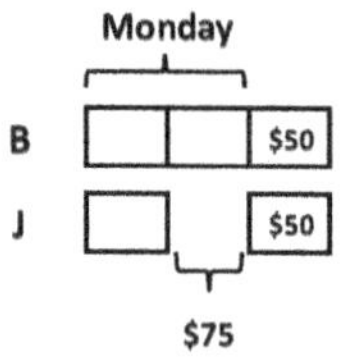

Challenge 11: Jack has 55 candies.

Challenge 12: Finn has 21 apples.

Challenge 13: There are 46 girls and 52 boys.

Challenge 14: The numbers are 155, 156 and 157.

Challenge 15: Kat started with \$40 and Bobby had \$20.

Challenge 16: Sarah scored 8 points.

Challenge 17: Bella has 38, Ron has 25 and Milo has 19 marbles.

Challenge 18: There are 181 fifth graders.

Fraction Frenzy Warm-Up 1 (p. 12)

1. $\frac{5}{6}$
2. $1\frac{2}{15}$
3. $\frac{7}{18}$
4. $4\frac{1}{4}$
5. $6\frac{1}{9}$
6. $6\frac{1}{3}$
7. $1\frac{7}{24}$
8. $4\frac{5}{12}$
9. $4\frac{1}{3}$
10. $\frac{1}{32}$
11. $2\frac{2}{5}$
12. 11
13. $\frac{1}{9}$
14. 40
15. $\frac{1}{16}$
16. $\frac{35}{12}$
17. $\frac{8}{3}$
18. $\frac{5}{8}$

Middle School Math Quest

Fraction Frenzy Warm-Up 2 (p. 13)

1. $\frac{5}{7}$
2. $1\frac{4}{15}$ or $\frac{19}{15}$
3. $\frac{1}{6}$
4. 6
5. $4\frac{2}{3}$
6. $3\frac{3}{4}$
7. $\frac{5}{6}$
8. $3\frac{3}{4}$
9. $2\frac{1}{2}$
10. $\frac{1}{5}$
11. 2
12. 13
13. $\frac{2}{15}$
14. 12
15. $\frac{1}{16}$
16. $3\frac{5}{6}$
17. 1
18. $\frac{5}{12}$

Fraction Frenzy, Super Strategies (pp. 14-18)

1. $13\frac{11}{15}$
2. $1\frac{7}{18}$
3. $23\frac{5}{8}$
4. $21\frac{7}{10}$
5. $9\frac{1}{6}$
6. $23\frac{9}{10}$
7. $\frac{361}{40}$ or $9\frac{1}{40}$
8. $\frac{72}{55}$ or $1\frac{17}{55}$
9. $\frac{133}{12}$ or $11\frac{1}{12}$
10. $6\frac{3}{10}$
11. $16\frac{1}{4}$
12. $93\frac{11}{12}$
13. $\frac{1}{2}$
14. 1
15. $\frac{35}{2}$
16. $\frac{9}{28}$
17. $\frac{36}{49}$
18. $13\frac{23}{27}$
19. 10
20. $\frac{6}{5}$
21. $\frac{22}{5}$
22. $5\frac{2}{15}$
23. $1\frac{3}{8}$
24. $1\frac{21}{124}$
25. $\frac{19}{27}$
26. $\frac{8}{9}$
27. $\frac{5}{8}$
28. $\frac{3}{4}$
29. $\frac{4}{5}$
30. $\frac{21}{109}$
31. $\frac{1}{11}$
32. $1\frac{3}{5}$
33. $5\frac{1}{22}$
34. $\frac{23}{34}$
35. 20
36. $\frac{5}{12}$

Fraction Frenzy, Advanced Practice (p. 19)

1. $\frac{1}{3}$
2. $\frac{21}{80}$
3. $1\frac{1}{3}$
4. $3\frac{13}{36}$
5. $62\frac{17}{24}$
6. $6\frac{13}{36}$
7. $5\frac{1}{2}$
8. $17\frac{11}{24}$
9. $44\frac{1}{2}$
10. 25
11. $1\frac{17}{18}$
12. $31\frac{7}{20}$
13. $39\frac{13}{20}$
14. $2\frac{11}{12}$
15. $78\frac{3}{10}$
16. $17\frac{19}{24}$
17. 60
18. $1\frac{19}{23}$
19. $\frac{1}{2}$
20. $5\frac{2}{3}$
21. $\frac{1}{5}$
22. 1
23. $5\frac{13}{18}$
24. $\frac{3}{28}$

Fraction Frenzy, Modeling and Visualization (pp. 20-21)

1. They need to make $23\frac{1}{3}$ quarts (or 5 gals and $3\frac{1}{3}$ quarts).

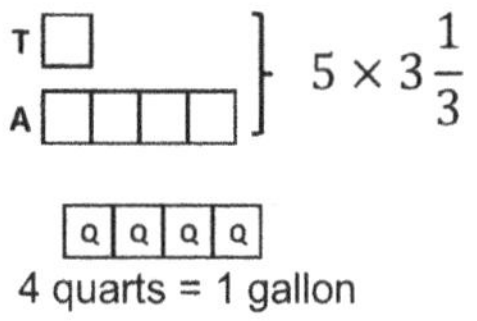

2. Emily sells 11 boxes in all.

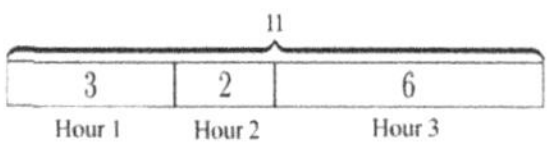

3. They bought 19 pairs of earrings in total.

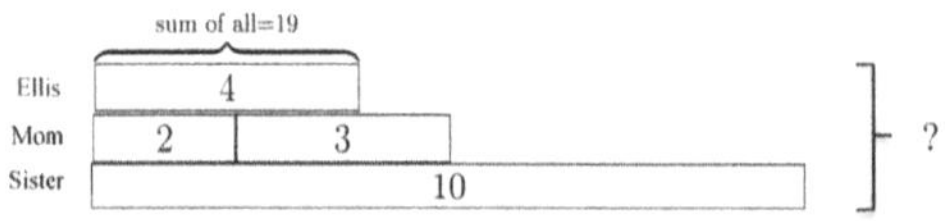

4. Steve has 10 black hats.

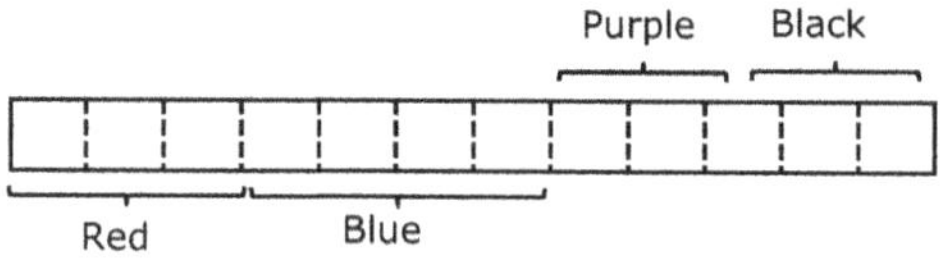

5. $8\frac{1}{3}$ acres are available for the stadium.

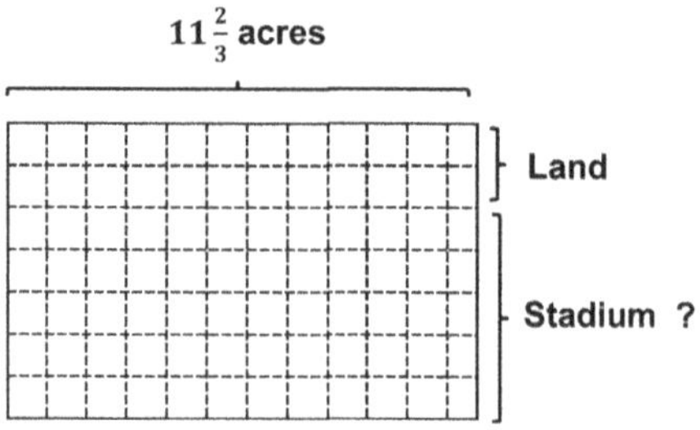

6. Peter started with 48 tickets.

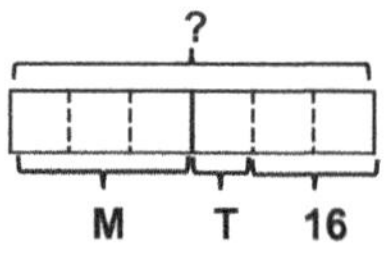

7. Example Model

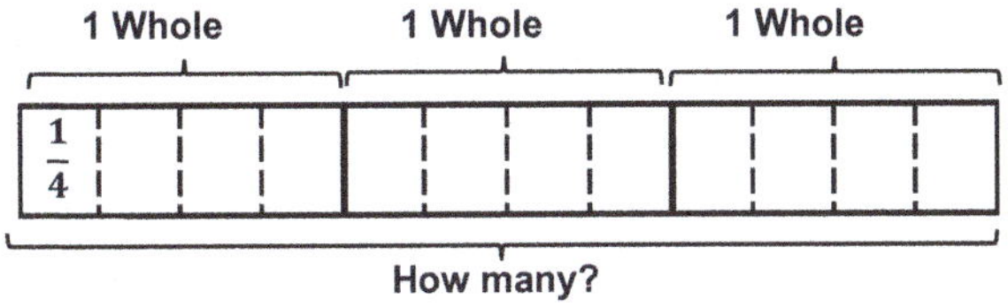

8. Example Model

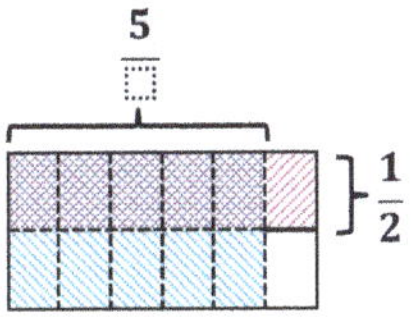

9. Example Model

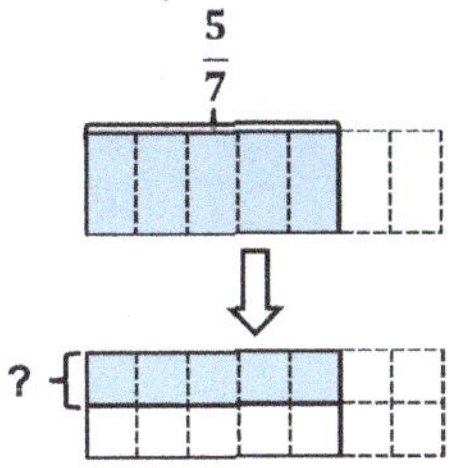

10. Sample Question: Susan has $2\frac{2}{3}$ lbs of candy that she wants to divide into smaller bags which hold $\frac{1}{6}$ lb of candy. How many smaller bags can she fill?

Sample Answer: Susan can fill 16 smaller bags with candy.

The answers for Q11 through Q19 might change if you include negative numbers. But, we talk about those later in the book (Integer Intrigue) so these answers assume only positive numbers.

11. Always true. When you divide a number by another number that is greater than 1, the answer is smaller than the original number.
Example: $6 \div \frac{3}{2} = 6 \times \frac{2}{3} = 4$

12. Sometimes. This is true when *n* is multiplied by a number greater than one. When *n* is multiplied by a number that is less than one, the answer will be smaller than *n*.
True Example: 3 x 4 = 12 False Example: $3 \times \frac{1}{3} = 1$

13. Sometimes. When you divide a whole number by a proper fraction (a number less than one), you are creating more groups than you originally had. The answer will be greater than the original number. But, when you divide a fraction by a fraction, you can get an answer greater than 1!
False Example: $10 \div \frac{1}{2} = 10 \times 2 = 20$ True Example: $\frac{2}{3} \div \frac{1}{2} = \frac{2}{3} \times 2 = \frac{4}{3}$

14. Sometimes. True example is $10 \div \frac{1}{4} = 10 \times 4 = 40$. False example is $10 \div \frac{4}{3} = 10 \times \frac{3}{4} = 7\frac{1}{2}$

15. Never true. $\frac{1}{2} \times 20 = 10$; $\frac{1}{2} \times \frac{1}{4} = \frac{1}{8}$

16. Always true. $\frac{3}{4} \times \frac{3}{8} = \frac{9}{32}$; $\frac{1}{5} \times \frac{2}{7} = \frac{2}{35}$

17. $\frac{5}{2} \times 1\frac{1}{3} = 3\frac{1}{3}$ and $3\frac{1}{3} > 1\frac{1}{3}$

18. $6 \div \frac{20}{3} = 6 \times \frac{3}{20} = \frac{9}{10}$

19. $1\frac{1}{3} \times \frac{1}{3} = \frac{1}{2}$

Decimals, Money Mania (p. 22)

1. $4.45
2. $11.05
3. $7.35
4. $12.00
5. $5.00
6. $7.85
7. $12.30
8. $42.50
9. $22.60

Decimals, Powerful Patterns (p. 24)

When You…	A Number By…	Move the Decimal Point….	Place(s) to the…	Is Answer Smaller or Larger Than Original Number?
Multiply	10	One	Right	Larger
Multiply	100	Two	Right	Larger
Multiply	0.1	One	Left	Smaller
Multiply	0.01	Two	Left	Smaller
Divide	10	One	Left	Smaller
Divide	100	Two	Left	Smaller
Divide	0.1	One	Right	Larger
Divide	0.01	Two	Right	Larger

Decimals, Decimal Discovery (p. 25)

1. 24; 0.24; 0.024

   Example Fraction Work: $0.6 \times 0.4 = \frac{6}{10} \times \frac{4}{10} = \frac{24}{100} =$ 0.24

2. 2,400; 24,000; 240,000
3. 0.24; 0.24; 24
4. 4; 0.4; 0.04

   Example Bar Model Work: Solve $0.24 \div 6$

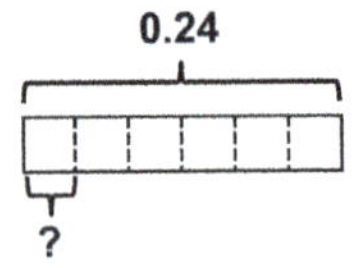

5. 40; 40; 4
6. 0.04; 0.004; 0.004
7. 60; 6; 0.6

   Example Sequence Work: Solve $2.4 \div 0.4$

   $24 \div 4 = 6$

   $2.4 \div 4 = 0.6$

   $2.4 \div 0.4 = 6$

8. 6,000; 6,000; 600,000
9. 180; 20; 180
10. 40; 8; 0.8
11. 9,000; 0.063; 0.07
12. 40; 36; 360

Order, Order (p. 26)

Group 1: 6.01, 6.0035, 6.003, 0.62, 0.6095

Group 2: $2\frac{1}{2}, 2\frac{1}{6}, 1\frac{4}{5}, 1\frac{2}{3}, \frac{3}{5}$

Group 3: $2\frac{7}{9}, 2\frac{1}{2}$, 2.46, 2.12, $1\frac{3}{5}$

Group 4: $2\frac{3}{4}, 2\frac{1}{3}$, 2.14, 2.102, 2.014

1. 37
2. 232
3. 51
4. 16
5. 27
6. $10\frac{7}{8}$
7. $\frac{5}{8}$
8. $1\frac{1}{4}$
9. $2\frac{1}{2}$
10. $3\frac{13}{16}$

Amazing Algebra (pp. 27-31)

1. $n - 3$
2. $2 \times y$
3. $x + 6$
4. $\frac{1}{2} \times z$
5. $\frac{1}{2} \times x - 5$
6. $3 \times b + 4$
7. 28
8. 20
9. 5
10. $4x + 1$
11. $4y + 5$
12. $3 + 14x$
13. $x + y + 1$
14. 0
15. $x = 10$
16. $x = 0$
17. $x = 40$
18. $x = 23$
19. $x = 15$
20. $x = 14$
21. $x = 419$
22. $x = 987$
23. $x = \frac{1}{8}$
24. $y = 45$
25. $x = \frac{17}{20}$
26. $n = 108$
27. $x = 12$
28. $x = 32$
29. $y = 5$
30. $x = 12$
31. $x = 297$
32. $x = \frac{1}{16}$
33. $y = 5$
34. $a = 4$
35. $b = 5$
36. $h = 172.5$

Algebraic Tug of War (pp. 32-36)

1. **4**, **8**
2. **6**
3. **6**, **1**
4. **2**, **5**, Three red people and one pink person.
5. **5**, **21**

6. **14**, **5**, **9**

7a. **6**

7b. **20**

7c. **5**. Right now, the right side is pulling with a force of 24. To pull at three times that force (72), they would need to add enough **Blues** to at least match a force of 48. So, they'd have to invite at least 10 more **Blues** to join them.

8a. **8**, **6**

8b. 13 people total. 13 red people, or 12 red and 1 blue.

9. **2**, **8**

10. **6**, **3**, She can't pull up the last vine on her own, you would need three Angelas.

Probability Palooza (pp. 37-39)

1. $\frac{5}{6}$

2. $\frac{1}{3}$

3. $\frac{1}{4}$

4. $\frac{3}{36}$

5. 36 Ways

Outcome Table for Rolling Two Dice

		Outcomes for Dice 2					
		1	2	3	4	5	6
Outcomes for Dice 1	1	(1,1)	(1,2)	(1,3)	(1,4)	(1,5)	(1,6)
	2	(2,1)	(2,2)	(2,3)	(2,4)	(2,5)	(2,6)
	3	(3,1)	(3,2)	(3,3)	(3,4)	(3,5)	(3,6)
	4	(4,1)	(4,2)	(4,3)	(4,4)	(4,5)	(4,6)
	5	(5,1)	(5,2)	(5,3)	(5,4)	(5,5)	(5,6)
	6	(6,1)	(6,2)	(6,3)	(6,4)	(6,5)	(6,6)

6. $\frac{1}{9}$

7. 15 Outcomes (see Sum Table)

Outcome Table for Sum of Two Dice

		Outcomes for Dice 2					
		1	2	3	4	5	6
Outcomes for Dice 1	1	2	3	4	5	6	7
	2	3	4	5	6	7	8
	3	4	5	6	7	8	9
	4	5	6	7	8	9	10
	5	6	7	8	9	10	11
	6	7	8	9	10	11	12

8. $\frac{5}{12}$

9. 1

10. $\frac{5}{18}$

11. $\frac{19}{37}$

12. 60 Ways

13. 9 x 9 x 8 x 7 x 6 = 27,216 code

Geometry Concepts (pp. 41-43)

1. q = 60°
2. p = 25°, q = 155°, r = 65°
3. q = 60°, Triangle: Equilateral
4. q = 80°, Triangle: Isosceles
5. q = 30°, Triangle: Scalene
6. 12 meters long
7. 8 millimeters
8. 50 x $1.93 = $96.50
9. A = 8 x 6 = 48 cm
The average of the bases is the same thing as $\frac{1}{2}\times(b_1+b_2)$.
10. $A=\frac{1}{2}\times 12\times 6=36$
11. $A=\frac{1}{2}\times(14+18)\times 11=$ 176
12. A = 9 x 7 = 63
13. SA =2(4x6 + 4x8 + 6x8) = 208
14. V = 4 x 6 x 8 = 192
15. 6 x 6 + 4 x 32 = 164
She needs 164 square inches of material.

Radical Ratios (pp. 44-46)

1. 4 squares to 6 circles
2. $\frac{3}{4}$
3. 3:6
4. 6 to 13
5. 18 girls, 12 boys
6. 1 to 2
7. 5 shirts
8. $\frac{2}{7}$
9. 3 to 2
10. 40 voters
11. $85
12. Greg is faster
13. $300
14. Car B
15. Box A

Percents Plus (pp. 47-51)

1. 90
2. 48
3. 45
4. 150
5. 60
6. 6
7. 60
8. 40
9. 60%
10. $133.\overline{33}\%$
11. 40%

FRACTION	DECIMAL	PERCENTAGE
$\frac{5}{8}$	0.625	62.5%
$\frac{1}{10}$	0.1	10%
$\frac{2}{3}$	$0.\overline{66}$	$66.\overline{66}\%$
$\frac{2}{5}$	0.4	40%
$\frac{1}{5}$	0.2	20%
$\frac{1}{20}$	0.05	5%
$\frac{6}{25}$	0.24	24%
$\frac{2}{25}$	0.08	8%
1	1.0	100%
$\frac{3}{4}$	0.75	75%
$\frac{3}{10}$	0.3	30%
$2\frac{3}{4}$	2.75	275%
$1\frac{1}{4}$	1.25	125%
$\frac{1}{3}$	$0.\overline{33}$	$33.\overline{33}\%$

FRACTION	DECIMAL	PERCENTAGE
$\frac{17}{20}$	0.85	85%
$\frac{99}{100}$	0.99	99%
$\frac{4}{5}$	0.8	80%
$\frac{1}{4}$	0.25	25%
$\frac{1}{8}$	0.125	12.5%
$1\frac{5}{6}$	$1.8\overline{33}$	$183.\overline{33}\%$
$\frac{19}{20}$	0.95	95%
$\frac{9}{20}$	0.45	45%
$\frac{1}{6}$	$0.1\overline{66}$	$16.\overline{66}\%$
$\frac{11}{20}$	0.55	55%
$\frac{7}{9}$	$0.\overline{77}$	$77.\overline{77}\%$
$\frac{9}{50}$	0.18	18%
$\frac{1}{100}$	0.01	1%
$\frac{9}{10}$	0.9	90%

12. 500%

13. 17

Domino Ratios (pp. 52-54)

1. This is just one possible answer. However, your solution must have 4 blue, 10 yellow, and 14 red squares.

2. Here's one answer. Make sure your top half has 4 red, 8, yellow, and 12 blue squares while your bottom half should have 12, red, 8 yellow, and 4 blue tiles.

3. Your solution may look different, but you should have 5 blue, 10 yellow, and 30 red squares.

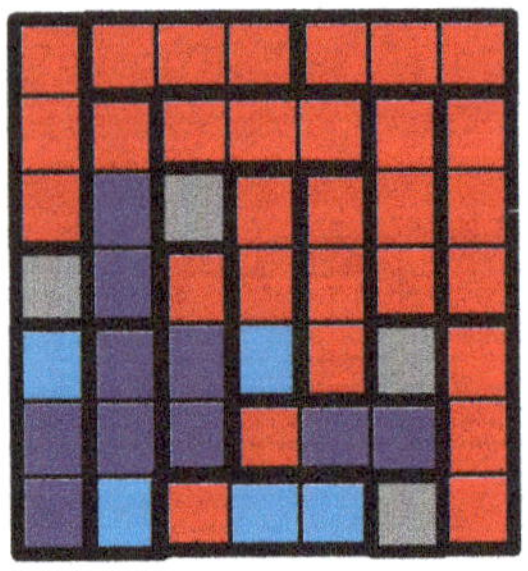

4. The maximum number of red spaces is 22. This solution has a 11:6:5 ratio of red:green:blue.

Intro to Exponents (pp. 55-56)

1. 16
2. 125
3. 81
4. 1,000
5. 1,331
6. 1
7. 1,024
8. 27
9. 16
10. 729
11. 1, 4, 9, 16, 25, 36, 49, 64, 81, 100, 121, 144, 169, 196, 225, 256, 289, 324, 361, 400
12. 1, 8, 27, 64, 125, 216, 343, 512, 729, 1000; $1 = 1^2$ *and* 1^3; $64 = 8^2$ *and* 4^3
13. 10 is the base

Prime Factorization (pp. 57-58)

1. $2^2 \times 3$
2. $2^2 \times 3 \times 7$
3. 2^6
4. $2^2 \times 3 \times 5$
5. 37
6. 3^4
7. 2×13
8. $2^3 \times 5^3$
9. 2^8
10. $2^2 \times 3^2$

Prime Factorization, GCF and LCM (pp. 59-60)

1. GCF =20; LCM = 240
2. GCF = 8; LCM = 840
3. GCF = 40; LCM = 80
4. GCF = 36; LCM = 144
5. GCF = 9; LCM = 351
6. GCF = 2; LCM = 128
7. GCF = 5; LCM = 75
8. GCF = 2; LCM = 684
9. GCF = 20; LCM = 400

Divisibility Rules (pp. 61-62)

NUMBER	2	3	4	6	9	15	18	24
60	Y	Y	Y	Y	N	Y	N	N
72	Y	Y	Y	Y	Y	N	Y	Y
144	Y	Y	Y	Y	Y	N	Y	Y
300	Y	Y	Y	Y	N	Y	N	N
360	Y	Y	Y	Y	Y	Y	Y	Y
163,536	Y	Y	Y	Y	N	N	N	Y
18,024	Y	Y	Y	Y	N	N	N	Y
32,303	N	N	N	N	N	N	N	N
4,664	Y	N	Y	N	N	N	N	N
165	N	Y	N	N	N	Y	N	N
3,280	Y	N	Y	N	N	N	N	N
15,824	Y	N	Y	N	N	N	N	N
162,150	Y	Y	N	Y	N	Y	N	N
2,940	Y	Y	Y	Y	N	Y	N	N
28,652	Y	N	Y	N	N	N	N	N
13,411	N	N	N	N	N	N	N	N
2,793	N	Y	N	N	N	N	N	N
148,264	Y	N	Y	N	N	N	N	N
19,782	Y	Y	N	Y	Y	N	Y	N
6,435	N	Y	N	N	Y	Y	N	N

Interesting Fractions (pp. 63-64)

1. $\frac{2}{3}$	2. $\frac{1}{4}$	3. $\frac{1}{3}$	4. $\frac{3}{2}$
5. 6	6. $\frac{13}{6}$	7. $\frac{1}{9}$	8. $\frac{5}{6}$
9. $\frac{1}{2}$	10. $\frac{8}{3}$	11. $-\frac{5}{4}$	12. $\frac{3}{2}$
13. $\frac{35}{54}$	14. $12\frac{3}{4}$	15. $-\frac{3}{7}$	16. $-164\frac{2}{3}$

Challenge: $\frac{5}{8}$

Integer Intrigue (pp. 65-71)

1. 2	11. -45 ft	21. −10	31. −1
2. −20	12. −690	22. −10	32. −27
3. 10	13. 145	23. −21	33. 16
4. 7	14. −543	24. −31	34. 4
5. 16	15. −170	25. −92	35. −21
6. −3	16. −8	26. 8	36. 2
7. $-3, 0, 1, \lvert-2\rvert, \lvert-3\rvert$	17. −241	27. 91	37. 33
8. $-10, -8, -7, -\lvert-4\rvert, \lvert-12\rvert$	18. −502	28. 0	38. 138
9. 8	19. −205	29. −1	39. 10
10. -2,000 ft	20. −391	30. 61	40. 1,162

Greens versus Aliens (pp. 72-76)

Challenge 1

Generation 4

Player Grid	🙂		
			👽
	🙂	🙂	
Sums Table	0	0	-1
	3	2	1
	1	0	0
Products Table	1	-1	-1
	1	-1	1
	1	-1	-1

Generation 5

Player Grid			👽
Sums Table	0	-1	0
	0	-1	-1
	0	0	0
Products Table	1	-1	1
	1	-1	-1
	1	1	1

Generation 6

Player Grid		👽	
		👽	👽
Sums Table	-2	-2	-3
	-2	-2	-2
	-1	-2	-2
Products Table	1	1	-1
	1	1	1
	-1	1	1

Challenge 2, Colony A

Generation 1

Player Grid	🙂		
		👽	
		🙂	
Sums Table	-1	0	-1
	1	2	0
	0	-1	0
Products Table	-1	-1	-1
	-1	1	-1
	-1	-1	-1

Generation 2

Player Grid			👽
	👽	👽	
Sums Table	-2	-3	-1
	-1	-2	-2
	-2	-2	-1
Products Table	1	-1	-1
	-1	1	1
	1	1	-1

Generation 3

Player Grid			
		👽	
			👽
Sums Table	-1	-1	-1
	-1	-1	-2
	-1	-2	-1
Products Table	-1	-1	-1
	-1	-1	1
	-1	1	-1

Challenge 2, Colony B

Generation 1

Player Grid	👽	👽	👽
	👽		👽
	👽	👽	👽
Sums Table	-2	-4	-2
	-4	-8	-4
	-2	-4	-2
Products Table	1	1	1
	1	1	1
	1	1	1

Generation 2

Player Grid	👽		👽
	👽		👽
Sums Table	0	-2	0
	-2	-4	-2
	0	-2	0
Products Table	1	1	1
	1	1	1
	1	1	1

Generation 3
COLONY DEATH!!!

Player Grid			
Sums Table			
Products Table			

Challenge 3, Colony C

Generation 1

Player Grid

Sums Table	0	1	1	1
	-1	0	0	1
	-2	-1	-2	0
	-1	-2	-2	-1
Product Table	1	1	1	1
	-1	-1	1	-1
	1	-1	1	1
	-1	1	1	-1

Generation 2

Player Grid

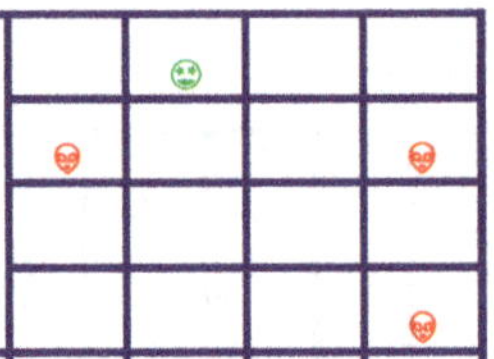

Sums Table	0	-1	0	-1
	1	0	0	0
	-1	-1	-2	-2
	0	0	-1	0
Product Table	-1	-1	-1	-1
	1	-1	-1	1
	-1	-1	1	1
	1	1	-1	1

Generation 3

Player Grid

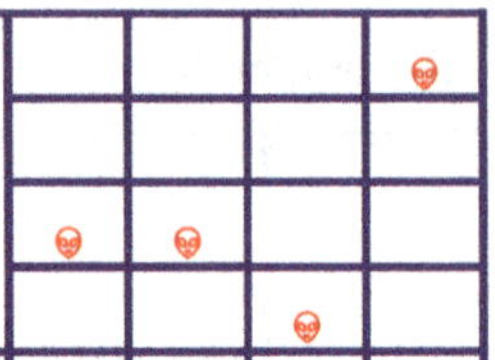

Sums Table	0	0	-1	0
	-2	-2	-2	-1
	-1	-2	-2	-1
	-2	-3	-1	-1
Product Table	1	1	-1	1
	1	1	1	-1
	-1	1	1	-1
	1	-1	-1	-1

Challenge 3, Colony D

Generation 1

Player Grid

Sums Table	0	0	1	1
	0	1	1	0
	1	1	0	1
	1	0	1	-1
Product Table	1	1	1	1
	1	1	-1	-1
	1	-1	1	-1
	1	-1	-1	1

Generation 2

Player Grid

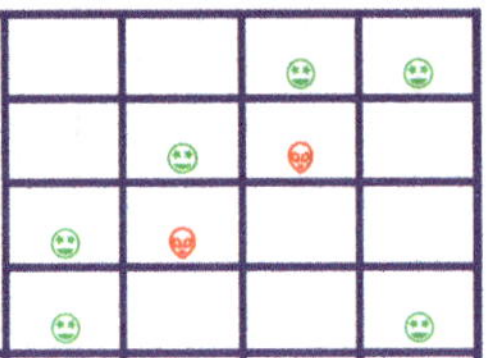

Sums Table	1	1	1	0
	1	0	2	1
	1	2	0	0
	0	1	0	0
Product Table	1	-1	-1	-1
	-1	1	-1	-1
	-1	-1	1	-1
	-1	-1	-1	1

Generation 3

Player Grid

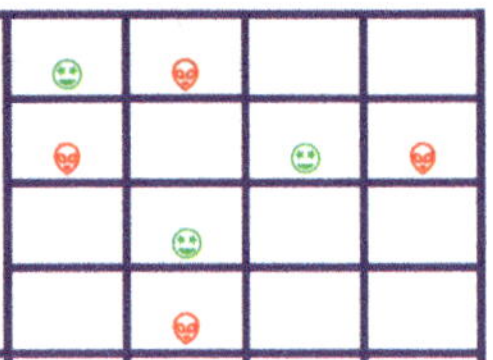

Sums Table	-2	1	-1	0
	1	1	-1	1
	-1	-1	0	0
	0	1	0	0
Product Table	1	-1	1	-1
	-1	1	1	1
	1	1	1	-1
	-1	1	-1	1

Middle School Math Quest

Challenge 3, Colony E

Generation 1

Player Grid		😀		
	😀	😀		
	👽	👽		
		👽		
Sums Table	3	2	2	0
	0	0	1	0
	0	0	-1	0
	-3	-2	-2	0
Product Table	1	1	1	1
	1	1	-1	1
	1	1	1	1
	-1	1	1	1

Generation 2

Player Grid		😀		
			👽	
			😀	
		👽		
Sums Table	1	-1	0	-1
	1	1	2	0
	-1	-1	-2	0
	-1	1	0	1
Product Table	1	-1	-1	-1
	1	-1	1	-1
	-1	1	1	-1
	-1	1	-1	1

Generation 3

Player Grid	😀			👽
	😀	👽	👽	
	👽	😀	😀	
	👽			😀
Sums Table	0	0	-3	-1
	0	2	0	-1
	0	-2	0	1
	0	0	3	1
Product Table	-1	1	-1	-1
	1	1	1	1
	1	1	1	-1
	-1	1	1	1

Challenge 4, Colonies F and G

Colony F Generation 1

Player Grid			😀	😀	
		👽		👽	👽
	👽				👽
			👽	👽	👽
Sums Table	0	1	1	1	1
	-1	0	0	0	-1
	-2	-1	-2	-2	-2
	-1	-3	-4	-6	-4
	-1	-2	-1	-3	-2
Product Table	1	1	1	1	1
	-1	-1	1	1	1
	1	-1	1	1	1
	-1	-1	1	1	1
	-1	1	-1	-1	1

Colony G Generation 1

Player Grid	😀			😀	
			👽	👽	😀
		👽	👽	👽	
	😀	👽	👽		
		😀			😀
Sums Table	0	0	-1	-1	0
	0	-2	-3	-1	-1
	-1	-3	-6	-3	-1
	-1	-1	-3	-2	0
	1	-1	-1	0	0
Product Table	1	-1	1	1	-1
	-1	-1	1	-1	1
	1	1	1	1	1
	1	-1	1	-1	-1
	-1	1	1	-1	1

Colony F Generation 2

Player Grid		😀			😀
	👽				😀
				👽	👽
	👽			😀	👽
Sums Table	0	-1	1	2	1
	1	0	0	0	-1
	-1	-1	-1	0	0
	-1	-1	0	-2	-2
	0	-1	1	-1	1
Product Table	-1	-1	1	1	1
	1	-1	-1	1	1
	-1	-1	-1	-1	-1
	-1	-1	-1	-1	-1
	1	-1	1	-1	1

Colony G Generation 2

Player Grid			😀		👽
			👽		
	😀	👽		👽	😀
			👽		
	👽		😀		
Sums Table	0	0	-1	-1	0
	0	0	-1	-1	-1
	-1	-1	-4	-1	-1
	-1	-1	-1	0	0
	0	-1	-1	0	0
Product Table	1	-1	-1	1	1
	-1	1	1	-1	1
	-1	1	1	1	-1
	1	-1	1	1	-1
	1	1	-1	-1	1

Colony F Generation 3

Player Grid			😀		
	👽	👽	👽		
	👽	👽			
		👽	😀		
Sums Table	0	1	0	1	0
	-2	-2	-1	0	0
	-3	-4	-2	-1	0
	-4	-4	-3	0	0
	-3	-1	-2	1	0
Product Table	1	1	1	1	1
	1	-1	1	-1	1
	-1	1	1	-1	1
	1	-1	1	-1	1
	-1	1	1	1	1

Colony G Generation 3

Player Grid				😀	
				👽	😀
	😀	👽			
		😀			
Sums Table	0	0	0	0	1
	0	0	0	2	0
	0	0	-2	0	0
	0	2	0	0	0
	1	0	0	0	0
Product Table	1	1	-1	-1	-1
	1	1	-1	1	-1
	-1	-1	1	-1	-1
	-1	1	-1	1	1
	-1	-1	-1	1	1

Positively Negativity (p. 77)

1. -22
2. $\frac{14}{3}$
3. 55
4. -10
5. -3
6. $-\frac{5}{16}$
7. 80
8. -4
9. 30
10. -1
11. 0.8
12. 16

Super Stats (pp. 78-83)

State 1

Generation 0

Colony	1		
		5	
Average	5	3	5
	3	1	5
	5	5	5

Median: 3

Generation 1

Colony			5
			5
	5	5	5
Average		5	5
	5	5	5
	5	5	5

Median: 5

Generation 2

Colony			5
			5
	5	5	5
Average		5	5
	5	5	5
	5	5	5

Median: 5

No mode for Generation 0. Generations 1 and 2 both have a mode of 5.

State 2

Generation 0

Colony		7	-4
	1		-15
Average	7	-4	7
	4	$-2\frac{3}{4}$	-4
		-7	

Median: -1.5

Generation 1

Colony	7		
	4		-4
		-7	
Average	4	$2\frac{1}{3}$	-4
	0	0	-7
	$-1\frac{1}{2}$	0	$-5\frac{1}{2}$

Median: 0

Generation 2

Colony		2	-4
			-6
Average	2	-4	2
	2	$-2\frac{2}{3}$	$-2\frac{2}{3}$
		-6	

Median: -4

No mode for any of generations of this state.

State 3

Generation 0

Colony	30	13	-30
	20	15	
Average	16	$8\frac{3}{4}$	14
	$19\frac{1}{3}$	$8\frac{1}{4}$	$-\frac{2}{3}$
	$17\frac{1}{2}$	$17\frac{1}{2}$	15

Median: 15

Generation 1

Colony			
	20		-1
	18	18	
Average	20	$9\frac{1}{2}$	-1
	18	$13\frac{3}{4}$	18
	19	$12\frac{1}{3}$	$8\frac{1}{2}$

Median: 18

Generation 2

Colony	20	10	-1
		14	
	18		9
Average	12	11	12
	$15\frac{1}{2}$	$11\frac{1}{5}$	8
	14	$13\frac{2}{3}$	14

Median: 12

No mode for Generations 0 or 2. Generation 1 has a mode of 18.

State 4

Generation 0

Colony				
		1		
			5	
Average	1	1	1	
	1	5	3	5
	1	3	1	5
		5	5	5

Median: 3

Generation 1

Colony	1	1	1	
	1			5
	1			5
		5	5	5
Average	1	1	3	3
	1	1	3	3
	3	3	5	5
	3	3	5	5

Median: 3

Generation 2

Colony	1	1		
	1	1		
			5	5
			5	5
Average	1	1	1	
	1	2	3	5
	1	3	4	5
		5	5	5

Median: 3

Generation 0 has no mode. Generations 1 and 2 have the same two modes, 1 and 5.

State 5

Generation 0

Colony	1		1	
		7		7
	1		1	
		7		
Average	7	3	7	4
	3	1	4	1
	7	4	7	4
	4	1	4	1

Median: 1

Generation 1

Colony		3		4
	3		4	
		4		4
	4		4	
Average	3	$3\frac{1}{2}$	$3\frac{2}{3}$	4
	$3\frac{1}{2}$	$3\frac{1}{2}$	$3\frac{3}{4}$	4
	$3\frac{2}{3}$	$3\frac{3}{4}$	4	4
	4	4	4	4

Median: 4

Generation 2, STEADY STATE

Colony		3		4
	3		4	
		4		4
	4		4	
Average	3	$3\frac{1}{2}$	$3\frac{2}{3}$	4
	$3\frac{1}{2}$	$3\frac{1}{2}$	$3\frac{3}{4}$	4
	$3\frac{2}{3}$	$3\frac{3}{4}$	4	4
	4	4	4	4

Median: 4

Generation 0 has a mode of 1. Generations 1 and 2 have a mode of 4.

State 6

Generation 0

Colony				
		-4		
		7		
Average	-4	-4	-4	
	-4		-4	
	$1\frac{1}{2}$	$1\frac{1}{2}$	$1\frac{1}{2}$	
	7		7	

Median: 1.5

Generation 1

Colony	-4	-4	-4	
	-4		-4	
	7		7	
Average	-4	-4	-4	-4
	-4	-4	-4	-4
	$1\frac{1}{2}$	$1\frac{1}{2}$	$1\frac{1}{2}$	$1\frac{1}{2}$
		7		7

Median: -4

Generation 2

Colony	-4	-4	-4	
	-4		-4	
	2	2	2	2
		7		7
Average	-4	-4	-4	-4
	-1	$-1\frac{3}{4}$	$-\frac{2}{5}$	-1
	$1\frac{2}{3}$	$\frac{3}{5}$	$2\frac{4}{5}$	$1\frac{2}{3}$
	$3\frac{2}{3}$	2	4	2

Median: 2

Generation 0 doesn't have a mode. Generation 1's mode is -4, the same as Generation 2's mode.

State 7

Generation 0

Colony	2		4	
		6		8
	-2		-4	
		-6		-8
Average	6	4	7	6
	2	0	$3\frac{1}{2}$	0
	0	$-1\frac{1}{2}$	0	$-1\frac{1}{3}$
	-4	-3	-6	-4

Median: 0

Generation 1

Colony		4		6
	2		4	
	-4		-6	
Average	3	3	$4\frac{2}{3}$	4
	4	$3\frac{1}{3}$	5	5
	-1	-1	-1	-1
		-5		-6

Median: 3

Generation 2

Colony		4		
			4	5
	-1	-1	-1	-1
		-5		-6
Average	4	4	$4\frac{1}{3}$	$4\frac{1}{2}$
	$\frac{2}{3}$	1	$1\frac{1}{5}$	$\frac{2}{3}$
	-3	$-\frac{3}{4}$	$-\frac{2}{3}$	$\frac{1}{2}$
	$-2\frac{1}{3}$	-1	$-2\frac{4}{5}$	-1

Median: -1

Generation 0 doesn't have a mode. Generation 1's mode is 4. Generation 2's mode is -1.

1. Using our formula, we have: $\frac{22+27+15+14+x}{5} = 20 \rightarrow \frac{78+x}{5} = 20 \rightarrow 78 + x = 100 \rightarrow x = 22$

2. $\frac{250+x}{6} = 60 \rightarrow x = 85$

3. 6

4. She scored 20 goals in total.

5. The average is 6 which is the median of the dataset. When groups of numbers are evenly spaced, we have an arithmetic sequence and the average will equal the median.

6. The average is 26, the same as our median.

7. The average is 51. This is still our median value. In this case, though, our median isn't one of our observations because we have an even number of numbers in our dataset.

8. High temp on Day 6 was -50.

9. We only have to calculate observation 5 and observation 6 to get our median. It is 18.

10. The third number is $7\frac{5}{12}$.

Cross Number (p. 84)

	1 3	5	2 1		3 9	8	4 3	
5 2	6		6 4	7 8	2		8 3	9 2
9			10 1	4	2			9
	11 8	12 9	9		13 4	14 4	15 8	
16 7	8	2				17 2	8	5
	18 9	7	19 5		20 3	4	2	
21 5			22 9	23 3	2			24 1
25 9	26 9		27 6	5	3		28 2	4
	29 5	0	3		30 5	4	1	

Secret Codes (p. 85)
Double Trouble: 653,421

Clue 1: We know it's a 6-digit number. Can't start with a 9 or a 0
Clue 2: Tens digit will be an even number; Ones digit will be 1-4
Clue 3: Hundreds digit is 4 or 8; Tens digit is 2 or 4; Ones digit is 1 or 2
Clue 5: Ten thousands digit is 2-9; Thousands digit is 0-7
Clue 6: Hundred thousands digit is 2, 4, 6 or 8; Thousands digit is 1-4; Ten thousands digit is 3-6
Clue 4: Work with remaining numbers to see which combos will give you the sums required

A Little Odd: 34,613

Clue 1: 5-digit number
Clue 2: Ones digit is odd
Clue 3: 1, 3 or 5 of the digits are odd; 0, 2 or 4 of the digits are even
Clue 4: Ten thousands digit is 1, 3, 5, 7 or 9
Clue 5: Thousands digit will be even and one more than the ones digit
Clue 6: Ones digit is 3 or 9; Tens digit is 1 or 3; Thousands digit is 4 which means ones digit is 3 and tens digit is 1
Clue 7: Ten thousands digit is 3
Clue 8: Since ones, tens and ten thousands digits are odd and thousands digit is even, the hundreds digit will be even. The only even multiple of 3 is 6.

Middle School Math Quest

Tower Math (p. 86)

Left Tower	Right Tower
4	
8	7
5	4
3	10
8	7
6	1
2	6
9	10

Triangle Challenges (p. 87)

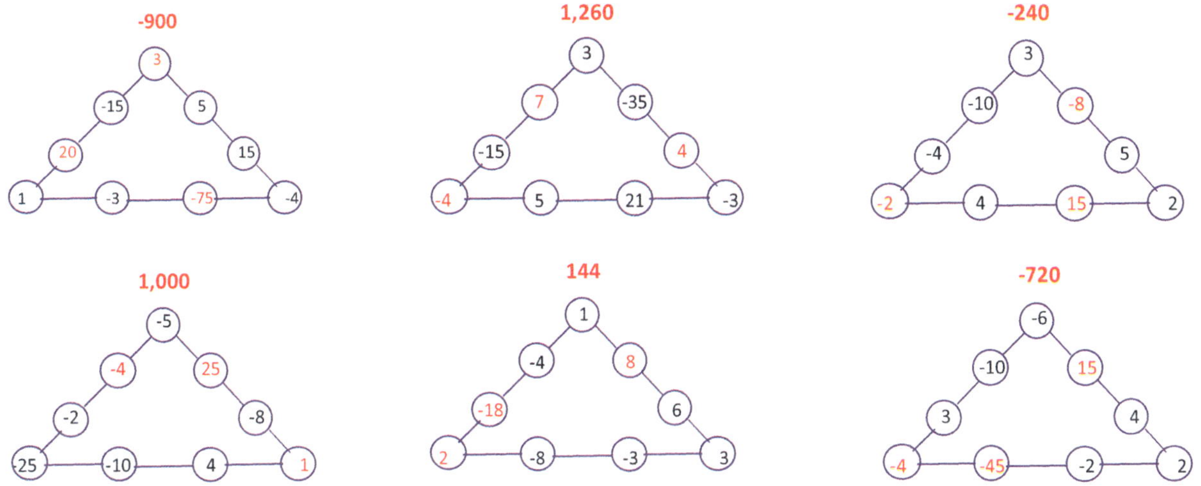

Multiplication and Division Quest (p. 88)

Challenge 1:	20	x	12	÷	2	x	10	÷	3	÷	8	=	50
Challenge 2:	34	÷	2	x	12	x	10	÷	4	÷	5	=	102
Challenge 3:	14	÷	7	x	12	÷	4	÷	3	x	12	=	24
Challenge 4:	40	x	8	x	8	x	12	÷	8	÷	10	=	384
Challenge 5:	30	x	10	÷	6	÷	2	÷	5	÷	5	=	1
Challenge 6:	26	x	10	÷	10	x	7	÷	13	÷	7	=	2
Challenge 7:	36	÷	2	÷	2	x	11	÷	3	÷	3	=	11
Challenge 8:	27	÷	3	x	7	x	9	÷	7	x	9	=	729
Challenge 9:	15	÷	3	x	8	x	6	÷	8	÷	3	=	10
Challenge 10:	17	x	6	÷	6	x	8	x	6	x	12	=	9,792

Divisibility Quizzle (p. 90)

Student	Favorite	Least Favorite
Kian	42	100
Lena	25	21
Maggie	360	2
Nate	77	3
Ollie	5	6

Made in United States
North Haven, CT
04 November 2026

21721415R00063